SPACE LAUNCH COMPLEX 10

Vandenberg's Cold War National Landmark

JOSEPH T. PAGE II

THE History PRESS

Published by The History Press
Charleston, SC
www.historypress.net

All cover images provided by the Vandenberg Space and Missile Technology Center.

First published 2016

Manufactured in the United States

ISBN 978.1.46713.631.0

Library of Congress Control Number: 2016943512

This book is dedicated to the men and women who worked out at SLC-10 and their families. They worked diligently under the veil of anonymity to provide support critical to national security…so this nation *shall not perish from the earth.*

In Memoriam
Eric G. Lemmon (1943–2016)
President of the Thor Association and 10th ADS Leader

CONTENTS

Forewords, by Daniel James Jr. and Richard C. Henry 7
Preface 9
Acknowledgements 11

1. SLC-10 at a Glance 15

Part I: The Past
2. The Mighty Thor 23
3. Project EMILY and the First Missileers 37
4. Nuclear Playground 53
5. America's Eyes in the Sky 75
6. The Fix to a Cloudy Problem 99
7. New Lease on Life 119

Part II: The Present
8. Vandenberg Space and Missile Heritage Center 127

Part III: The Future
9. SAMTEC and the Thor Historic District 137

Appendix 1: Overview of SLC-10 Facilities 147
Appendix 2: Launches from SLC-10 157
Appendix 3: Thor Specifications 161

Appendix 4: Lineage and Honors 163
Acronyms 167
Notes 171
Bibliography 179
Index 187
About the Author 191

FOREWORDS

070030Z NOV 75
UNCLASSIFIED
Subj: Favorable Communication

1. I commend 10th Aerospace Defense Squadron for your dedication, hardwork, and professionalism in support of the Army Ballistic Missile Defense Test Target Program (BMDTTP). Your outstanding performance contributed significantly to the successful accomplishment of the army's program objectives, so vital for our nation's future ballistic missile defense. Your contribution to the BMDTTP is even more praiseworthy when it is realized that you were simultaneously conducting the Category II testing of the new Defense Meteorological Satellite.
2. These achievements, coupled with your perfect record of 33 successful missile and space launches without a failure, clearly demonstrate the value and flexibility of an all military space launch organization. Please convey my personal congratulations and thanks to all personnel for a job well done.

Signed,

DANIEL JAMES, JR., General, USAF
Commander-in-Chief, Aerospace Defense Command

142100Z NOV 75
UNCLASSIFIED
Subj: Favorable Correspondence
1. Congratulations to your people at the Tenth Aerospace Defense Squadron, Vandenberg AFB, CA and to the crew deployed to Johnston Atoll for the second consecutive 100 percent successful launch of a Thor missile. From Johnston Island on 5 Nov 75. This totally successful two-launch series has resulted in the highest quality data acquired in the US Army Ballistic Missile Defense Test Target Program.
2. The professionalism, training, dedication, and teamwork of the crew are particularly praiseworthy. Please pass my personal congratulations and warm thanks for a job well done.

Signed,

RICHARD C. HENRY, Major General, USAF
Vice Commander, SAMSO

PREFACE

My first experience at Space Launch Complex 10 (SLC-10) was as a young lieutenant at Vandenberg Air Force Base, awaiting a missile combat crew training course. Six newly minted officers were given seemingly random directions by their supervisor to "drive toward the ocean" to find the Heritage Center. Seeing bits and pieces of broken rockets and missiles strewn about, I thought I knew what to expect—typical "vanilla" Air Force space history, watered down and politically palatable for mass consumption. In other words…boring. When I stepped out of my car and met Jay Prichard outside his office for the first time, I knew instantly I was dead wrong.

His fierce questions about our history knowledge and vibrancy of his presentation shook me to the core. Jay told us we were standing near the sites of some of the first operational launches in the American rocket and missile program. As a future missileer, this showed me the origins of the "can-do" spirit of the early missileers. For the future space launch officers next to me, this wasn't history; this was "Rockets 101, the preview" since the lineage of technology inside their Delta II, Titan and Atlas space launch boosters came directly from missiles launched at sites like these scattered around the base. I realized the distance from historic ideas to concrete "stuff" to see and experience was nowhere closer than at SLC-10.

Almost a decade after first meeting him, I was honored to befriend Jay and learn more about the missions performed at SLC-10, as well as other sites on base. Stories of airmen dropping cranes on missile-ready airframes and unofficial site tours with seductive Cuban spies were enough to entertain and inform the general public. But to feed my desire to know more, Jay provided food for thought not just about "what" was done at SLC-10 but also "how." I heard tales of true work ethic, punctuated by men and women willing to do their best to get the mission done, and of program veterans who never violated their oaths of silence to their country or service, even in the twilight of their lives.

One anecdote underpins the reason for this book. During one of my lunchtime visits to SLC-10, I brought Jay a declassified monograph I had dug out of my garage. I met him on the West Pad giving a tour to a group of ten individuals, a family, I later learned. Interrupting briefly, I said I couldn't stay long and gave the documents to him. The widest grin came over his face. "Wait a minute," he said. He walked over to an older gentleman and showed him the document. A quizzical look appeared on the man's face, and he responded with, "They declassified it?" I left, not understanding why Jay wanted me to wait.

I discovered the next day that the older gentleman was one of the original members of the Program 437 team. He had never told his family about his experiences on Johnston Island (JI) or at SLC-10. The children and grandchildren present that day finally learned what their patriarch did during his time in service. They might have never learned about that part of his life, nor would I, if not for a curious quirk of fate during that summer afternoon.

A bitter soul once remarked that all SLC-10 had to offer were "bloviated, fanciful stories of days gone by." I think the men and women who worked there, along with their friends and family, would vehemently disagree.

Any errors in fact or errors of omission are unintentional and mine alone. While the story of SLC-10 and its people may stretch beyond these words, I take full responsibility for the limited scope of this book.

ACKNOWLEDGEMENTS

I owe my eternal gratitude to Donald "Jay" Prichard, the Vandenberg Space and Missile Technology Center's executive director for over two decades. He has been the heart and soul of SLC-10 for as long as most people can remember. Jay helped pull back the layers of historical strata surrounding the base and helped me see beyond the statistical data and distilled sentences in the annual histories. Though the full story of SLC-10 may never be known, Jay has most volumes of that tale memorized. I must also thank Jay's constant canine companion and Mu-Q mooch, Sage. Only she and I know how many burgers we've shared.

Eternal love and cookies to "The Museum Underground," Esther and Natalie, for being all-around great sports and amazing sources of comic relief ("You know what's cool? Soccer!"). You don't ever realize how far down the rabbit hole you've fallen until an anthropologist and bioenvironmental technician, both civilians with no prior AF influence, tell you how screwed up your daily reality is.

Thanks go out to Colonel John Yocum (retired) and his son Rollan "Yoke" Yocum for providing perspective about the work done at SLC-10. Colonel Yocum is still "keeping it real" by forgoing e-mail and communicating via snail mail. Yoke, a hardened SAC warrior and Space Jedi Knight in his own right, has put up with me on and off since 2004.

Kudos to Ted Molczan, heralded as "The Guy" who can track satellites with paper, pencil and a calculator. His article on 437AP was eye-opening and very timely during the writing of this book. I'm sure he has many interesting stories to tell; perhaps one day I can hear more of them.

Special thanks go to Mr. Steve Cooke and Mr. John Boyes. These two gentlemen have been a great source of knowledge about Thor in the United Kingdom. Without their assistance on Project EMILY, the story would have been incomplete. I have not met them in person, but I do feel kinship as part of the family keeping the Thor story alive.

Thanks to Ms. Shawn Riem, 30th Space Wing historian, for providing key chronologies from Vandenberg's Cold War years. She's been subjected to my FOIA and requests for information since I met her in 2014. Peeking into 30th Space Wing archives would not have been possible without her help and encouragement.

Mr. Eric Lemmon, president of the Thor Association, provided excellent correspondence about life on Johnston Island during Program 437. While Eric is a source of incredible knowledge about the entirety of the Thor program, I scoped my questions due to his limited availability. I guarantee Eric's got a lot more to tell!

Thanks to Wayne Eleazer for his perspectives on the Thor shutdown. Wayne's articles on "The Space Review" about programs at Vandenberg offered interesting views on what the official annual histories leave out.

Peter Hunter and his collection of Thor photographs underpin most of this book's subject material. I have only heard stories of Peter's photo hunting exploits in the archives of the U.S. Air Force. I am grateful that he took the time to record and preserve the visual history of the Thor missile through the famed "Peter Hunter collection." I am truly indebted to him for conquering this. Perhaps one day we will meet.

As an unsung part of my writing and research process, I always give a shout-out to my baristas who keep me in iced caffeine goodness. Since part of this manuscript was written during my visits to Vandenberg, I must acknowledge the Santa Maria Starbucks team of Kiesha, Taylor, Paul and Karrie. My main coffee team in Albuquerque works at Satellite Coffee. While wholly apropos considering the book's subject material, Satellite Coffee makes a really good iced white mocha that fueled this adventure. Nods of thanks to Marcus, Avery, Hunter, Anthony and Bekah. If you ever want to get my undivided attention, an iced white mocha will do the trick.

To my editors at The History Press, Megan Laddusaw and Hilary Parrish, thank you for taking a chance on this project and all of your feedback, suggestions and edits.

Finally, I'd like to thank my family. To my wife, Kim, for her support during this book project. Hugs and kisses to my children—one day, you'll realize how important it is to remember the fabric of the past, from your own family or some greater venture. To my parents, Joe and Kathleen. I've said it before, in other words, but I will say it again: thank you for raising me to be curious to the point of insanity.

Chapter 1

SLC-10 AT A GLANCE

Space Launch Complex 10 (SLC-10) is one of three Thor launch complexes at Vandenberg Air Force Base (AFB), California. Originally built to support SM-75 Thor Intermediate Range Ballistic Missile (IRBM) training for Royal Air Force (RAF) missile launch crew members ("missileers"), the site has seen many changes during its decades of existence. After the RAF training program wound down in 1962, the Air Force took the facilities and equipment from SLC-10 and relocated them to Johnston Island (JI) to support U.S. nuclear testing efforts.[1]

SLC-10 consisted of three launch pads, two launch control centers and associated launch infrastructure, most of which still exists today. The three launch pads have had multiple names over the decades, illustrated here for the reader's clarification:

- West Pad: Complex 75-2-6, LE-6, SLC-10W
- East Pad: Complex 75-2-7, LE-7, SLC-10E
- Launch Emplacement 8: Complex 75-2-8, LE-8, 4300 B-6

In 1963, SLC-10 was rebuilt to support space launches, using equipment and facilities returning from the Project EMILY deployment of the Thor IRBM system to the United Kingdom. SLC-10 East (SLC-10E) supported training crew members for Program 437, America's first anti-satellite weapon. This weapon system was designed to hold hostile foreign satellites at risk, with two nuclear-tipped Thor vehicles on standby at JI. Initially a

A crew inside the SLC-10 Blockhouse prepares for a Defense Meteorological Support Program (DMSP) launch. *Courtesy of Vandenberg Space and Missile Technology Center.*

classified effort, President Lyndon B. Johnson revealed the program when confronting critics on the threat of Soviet space weapons.[2]

At the same time, SLC-10 West (SLC-10W) supported another critical space effort by launching Defense Meteorological Support Program (DMSP) satellites. These weather satellites were developed to assist a classified "black" program called CORONA, the world's first photoreconnaissance ("spy") satellites. Photographic film from early CORONA launches showed massive amounts of cloud cover over their targets in the Soviet Union. With DMSP, cloud cover could be predicted, allowing the cameras to avoid periods of dense cloud coverage and maximizing the intelligence "take" for the mission. DMSP information later trickled out of its classified world and began assisting U.S. military planners for a variety of missions. DMSP launches from SLC-10W began in 1965 and continued until 1980.[3]

A Thor IRBM, complete with British Royal Air Force roundel, lifts off from Launch Emplacement 8 (LE-8) on December 13, 1960. *Courtesy of Vandenberg Space and Missile Technology Center.*

After SLC-10 was deactivated in 1981, the site sat dormant until 1984, when the National Park Service began its search for relevant installations that played an important part in aerospace history. SLC-10 was recommended for inclusion in the National Register of Historic Places as "the best surviving example of a launch complex built in the 1950s at the beginning of the American effort to explore space."[4] The site was dedicated by the Department of the Interior as a National Historic Landmark on July 18, 1986.[5]

Today, the Vandenberg Space and Missile Technology Center at SLC-10 interprets the evolution of missile and space activity at Vandenberg from the beginning of the Cold War through current developments in military, civil, commercial and scientific space endeavors.[6]

Why You Should Care

Aside from being a really cool place to visit with a stunning view of the Pacific Ocean, SLC-10 is a reminder of the many noble endeavors in our past and the men and women who accomplished them. The support given to many launch programs (Thor IRBM, Program 437, DMSP) shows the versatility and forethought provided by the Thor vehicle designers—good ole American ingenuity in the face of uncertainty. The remnants of launch technology that we would consider "old" are still a generation or two beyond what some other countries have fielded in their rockets, a testament to American science and engineering. And finally, the stories stemming from the people of SLC-10 highlight astounding successes and equally momentous failures and are more than enough to inspire and motivate the next generation of rocket scientists and engineers…and who knows where *they* will take us?

The Thor Program: A Legacy of "Firsts"[7]

- First operational ballistic missile system in the free world
- First missile to be launched from Vandenberg Air Force Base (December 16, 1958)
- First booster to launch a spacecraft into polar orbit (Discoverer II, April 13, 1959)

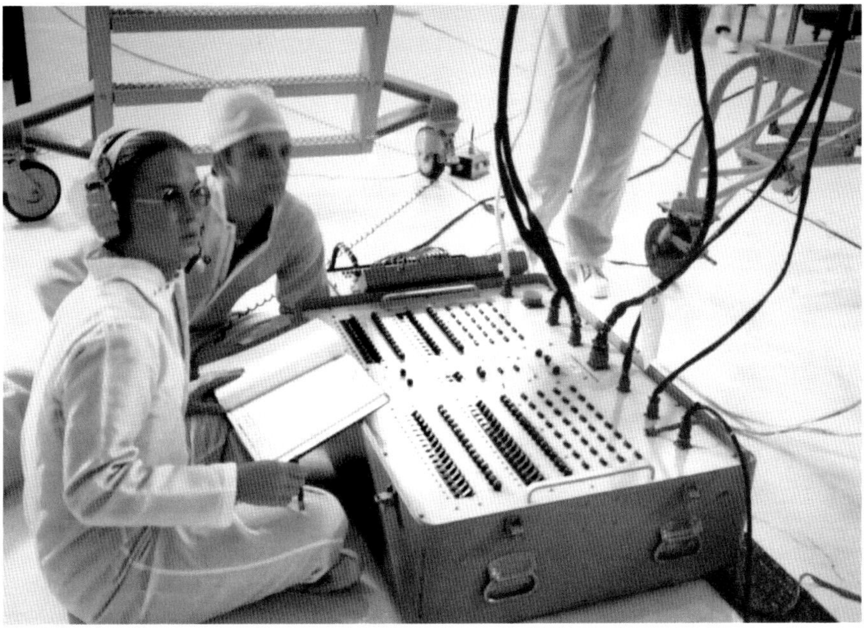

Sergeant Susan Mills, one of the first female launch crew members on a United States missile system, operates DMSP diagnostic equipment within the SLC-10 West Pad shelter. *Courtesy of Vandenberg Space and Missile Technology Center.*

- First booster to launch a payload recovered from orbit (Discoverer XIII, August 10, 1960)
- First booster to launch a communications satellite (Courier 1B, October 4, 1960)
- First booster to launch a meteorological ("weather") satellite (TIROS-1, April 1, 1960)
- First booster to launch a navigation satellite (Transit-1B, April 13, 1960)
- First long-range vehicle to record one hundred, two hundred, three hundred and four hundred launches
- First missile system to have female launch crew members[8]

PART I

THE PAST

Chapter 2

THE MIGHTY THOR

Development and Testing

The Thor Intermediate Range Ballistic Missile (IRBM) weapon system (also known as SM-75 and WS-315A) came about during planning for the logical follow-on to tactical missiles such as the Mace and Matador. Dr. James Killian released his eponymous "Killian Report" in Februrary 1955 to the National Security Council,[9] recommending the United States Air Force (USAF) develop a ballistic missile for land and ship launching with an inherent range of roughly 1,500 miles. The challenges within the longer-range intercontinental ballistic missile (ICBM) program were daunting. The problems behind a shorter-range system, however, were relatively less complicated, and the reasoning stood that the system could be finished sooner.

As early as 1951, during the development of the Matador missile, engineers at the Wright Air Development Center (WADC) showed that a ballistic missile would be literally invulnerable to interception while still achieving the required range and destruction required. Early goals within the ballistic missile program desired a range of between eight hundred and one thousand miles while carrying a payload of 3,000 pounds of explosives (1,500-pound payload for longer ranges). A Circular Error Probable (CEP) of 6,000 feet was achievable; however, a 1,500-foot CEP was desired.

The first Thor launch from Vandenberg Air Force Base, code named TUNE UP, lifts off on December 16, 1958. *Courtesy of Vandenberg Space and Missile Technology Center.*

CEP is a fancy term for a measure of a weapon system's precision. The measurement is the radius of an imaginary circle where 50 percent (one-half) of the weapons will land. In general, the smaller the CEP, the more precise the weapon. Newer ballistic weapons, such as the Minuteman, have CEP measurements around three hundred to six hundred *feet*. This becomes a bit of dark humor when considering the effects of nuclear weapons though. "Close enough for government work," indeed!

A design competition for this missile was initiated by Air Research and Development Command (ARDC); four contractors were selected to perform design studies (Glenn L. Martin, Douglas Aircraft Company, Lockheed Aircraft and General Electric).[10] While this design study was being accomplished, the deputy chief of staff for development, Lieutenant General Donald L. Putt, directed ARDC commander Lieutenant General Thomas S. Power to initiate a design *competition*. This was out of the ordinary during normal weapon system acquisition since the process usually awaited completion of the design study phase; however, the high priority of the program necessitated the move.

Speed was of the essence due to machinations in the global political climate. By 1955, the Soviet Union had operated ballistic missiles in the six-hundred- to eight-hundred-mile range. Its rapidly improving technology would ensure doubling of the range within a few years. The National Security Council recognized the psychological threat these terror weapons would give the European nations. It was of the utmost urgency to complete a weapon soon to counter their advantage.[11]

During November 1955, Secretary of Defense (SECDEF) Charles Wilson directed the USAF, Army and Navy to start efforts toward developing intermediate range ballistic missiles. While the USAF's would-be land-based, the joint Army/Navy program would provide a ship-based capability, as well as an alternative to the USAF's project.[12] Urgency would rear its ugly head once again, forcing Brigadier General Bernard A. Schriever, head of the Western Development Division, to ensure that all requests for proposals understood that the IRBM program would continue at the maximum rate technology would permit. (As proof of the importance of the IRBM program, and displaying complete faith in his ability to deliver, Schriever was promoted to major general rank in December 1955.)[13]

The U.S. government's official stamp of approval came on December 1, 1955, when President Dwight D. Eisenhower gave the Thor and Jupiter IRBM programs a rating of "DX," the highest national priority.[14] This placed the

IRBM programs on the same footing as the ICBM developmental programs (Atlas and Titan). While the programs were conducted concurrently, performance parameters for IRBMs were less restrictive than their ICBM cousins, allowing the Thor and Jupiter missiles to progress toward operational usefulness and field deployment more rapidly.[15]

The Douglas Aircraft Company received the contract to develop the WS-315A weapon system as the program's prime contractor on December 28, 1955.[16] The WS-315A designation encompassed the entire system, whereas the XSM-75 moniker was for the airframe. The name Thor came from the Norse god of thunder.[17] The development period was short—only thirteen months after the ink had dried on the contract to first launch—due in part to its simple design and concurrency concept with ICBM research. The linkage between the missile systems was quoted in an Air Force history of Thor:

> *The Air Force Ballistic Missile Division (formerly the Western Development Division), Major General Bernard A. Schriever, Commanding, has the entire management responsibility for the Air Force ballistic missile program. The Air Force Ballistic Missile Division is the management member of a three-part team. The second member is the Ballistic Missile Office of the Air Materiel Command, which exercises all contracting and procurement responsibilities for the program. The Guided Missile Research Division of the Ramo-Woolridge Corportation, the third member of the organization, provides technical direction and scientific supervision for the program, which now includes three inter-related weapon systems, Atlas, Titan and Thor.*[18]

With the U.S. desire to provide deterrence measures in a rapid manner, the design of Thor was kept simple—an aft section with the rocket propulsion, two propellant tanks (RP-1 kerosene and liquid oxygen) and the forward section containing the guidance and warhead.[19] The Douglas team had completed the engineering design and characteristics of Thor by the end of July 1956, while the manufacturing plant sprang up to fabricate the airframes. Concurrency was also used *within* the Thor program to save time. Engineering tests were being performed on the airframe as subcontractors were designing ground aerospace equipment and support structures to hold the missile erect before firing.[20]

The missile's final dimensions were sixty-three feet long and eight feet in diameter through the constant portion of its cylindrical sections. The gross weight of the missile was approximately 109,800 pounds, of which 98,000 were propellants. The forward section of the missile was tapered

As Luck Would Have It

During the race to build the IRBM, the Army's braintrust at Redstone Arsenal lost an unknowingly key figure: Adolf Thiel, a German engineer and veteran of the Nazis' V-2 program.[22] After Operation PAPERCLIP brought the Germans to the United States, Thiel worked at Redstone Arsenal for a few years on the Jupiter program. When his contract was up, he was lured to the West Coast to work on Thor. Along the way, Thiel had taken duplicates of the IRBM studies to California. The coincidences with both missile designs were not incidental. Both missiles had the exact same dimentions of diameter and fully fueled weight of 110,000 pounds, with Thor topping out at 64.8 feet over Jupiter's approximate 60.0-foot height. History would record Jupiter's main contribution as the "sacrificial lamb" for President John F. Kennedy's negotiations during the Cuban Missile Crisis, while the Thor program went on as the backbone of U.S. space launch capability into the twenty-first century.

for engagement with a separable reentry vehicle, containing a warhead capable of going 300 to 1,500 nautical miles.[21] The aluminum body was welded together, with internal stiffening to strengthen it throughout flight. The fuel tanks were about six millimeters thick, with internal baffles to prevent sloshing during flight. Additional parts of the body used skin-and-stringer construction, fastening on the aluminium skin to thin strips of material, like an internal metal skeleton. Access panels along the fuselage allowed external access to engage the system's innards. The rear of the airframe held the Rocketdyne MB-3 engine on gimbals so the thrust chamber could be swiveled to provide pitch and yaw. Roll control used the same turbo pumps as the main engine, with two 4.45 kN engines located diametricly opposite each other along the outer edge of the base plate.

The next year brought significant progress to the program. By March 1956, engine test facilities were being constructed at Edwards AFB, California, while launch facilities were being built at Patrick AFB, Florida.[23] Launch Complex 17 and 18 were used for the flight test program; each complex had one blockhouse controlling two launch pads. Complex 17 was used for "wet" testing, for tie-down testing where the missile was physically constrained to the pad. One "dry" pad at Complex 18 was used for flight launches only; the other pad and half of the blockhouse were dedicated to the Navy's Vanguard program.[24]

Teething problems inherent in any complex system reared its ugly head early during Thor's development. Critics of the concurrency concept had a field day with the early failures in the program. An iterative process

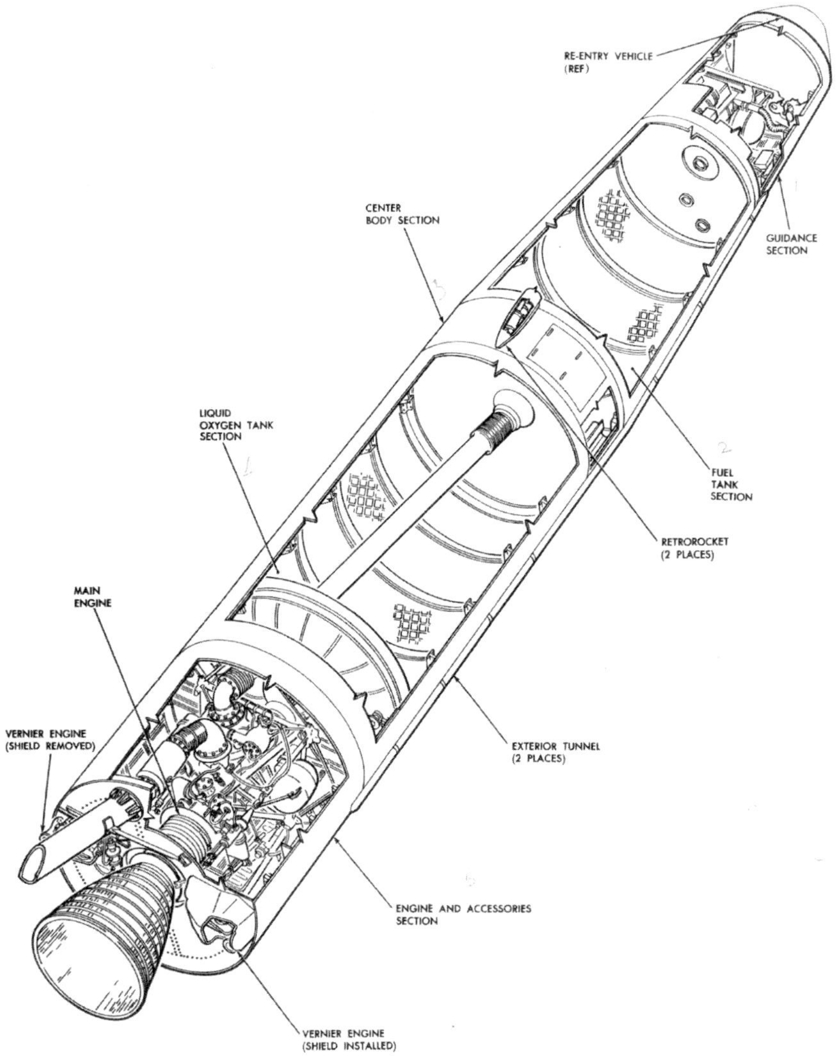

Thor Airframe

An interior cut-away of a Thor IRBM displays its deceptively simple design: warhead, guidance, liquid oxygen tank, fuel tank and rocket engine—key characteristics to its longevity and phenomenal launch success rate. *Courtesy of the United States Air Force.*

was undertaken to work toward an operational weapon system by using designations for each series of test launches.

The first missile, designated serial number 101, arrived at Patrick AFB on October 18, 1956, merely nine and a half months after the program's initiation. At the designated launch time, Thor No. 101 rose almost two feet off the ground before exploding into a fireball. Author Neil Sheehan explains how the post mortem of the accident showed a probable cause: "[A public relations film of the launch] showed two technicians...pulling a hose that was to be used to fill the oxidizer tank of the Thor with liquid oxygen."[25]

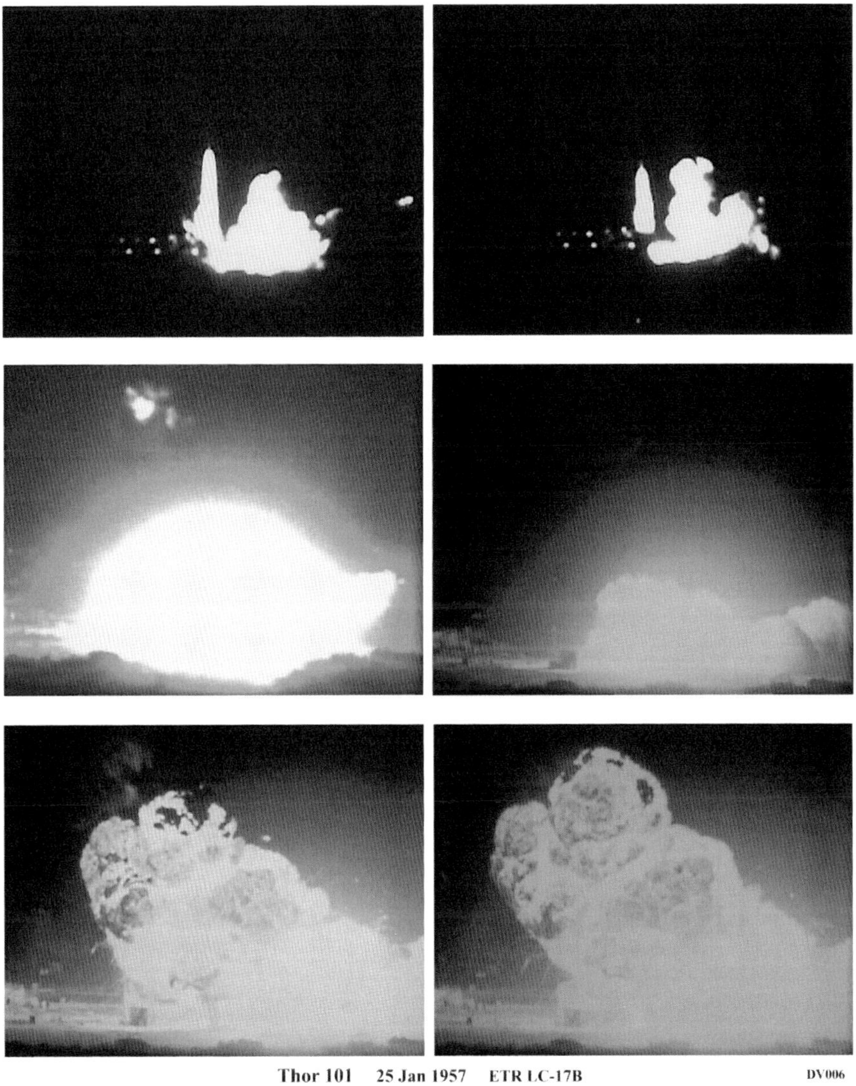

A time-lapse collage shows the destruction of the first Thor missile, serial number 101, at Cape Canaveral on January 25, 1957. *Courtesy of the United States Air Force.*

The technicians pulled the liquid oxygen (LOX) hose along the ground through sand before fitting it to the missile. After performing a similar experiment under controlled conditions at a Rocketdyne test facility (and creating a similar explosion!), it was determined that minute grains of sand had contaminated the fueling system and exploded when the LOX was ignited. Sheehan recounts the aftermath: "In the future, all hoses, valves, and other connections were kept in a state of pristine cleanliness that became known in the liquid-fueled rocket business as 'LOX-clean.'"[26]

The launch of the second Thor, no. 102, had the curious distinction of being "right" while also being "wrong," in a manner of speaking. Lift-off of the missile from the launch pad was achieved without incident. However, after thirty-two seconds of flight over the Atlantic Ocean, the range safety officer (RSO) pressed the missile destruct button, to the alarm of the viewing audience. When questioned about the missile's destruction, the officer claimed the missile was heading inland toward populated areas. One of his instruments had been installed incorrectly—backward, in fact.[27] So the flight path of Thor 102 soaring over the ocean looked like a stray missile fired inland to the RSO from his console inside the blockhouse. The Air Force finally had a missile that could fly, albeit only if given the chance.

The next missile, Thor No. 103, met an inglorious end while still waiting on the launch pad. Worker fatigue was cited as the cause. The testing took place over a twenty-four-hour period (May 21–22, 1957), with many stops and starts for troubleshooting. During the long wait times between countdown re-initiation, a large volume of LOX boiled off. During the subsequent LOX loading to refill the tank, a tired worker missed the gauge indications that showed the tank pressure had grossly exceeded limits. Furthermore, the destruction of No. 103 severely damaged the launch pad and surrounding equipment, adding more delays to an already struggling program.

Thor No. 104 also met a fiery end on August 30, 1957, but this time it was in flight over the Atlantic Ocean. The missile flew true for a little over a minute and a half before being ripped apart by aerodynamic stresses. A bad electrical signal had triggered the yaw actuator to turn the missile abruptly.

The launch of missile No. 105 on September 20, 1957, would finally give the USAF/Douglas team its first successful test of the Thor weapon system from launch to splashdown. The missile flew down the Atlantic, performed engine cut-off at the appropriate time and lofted its dummy warhead beyond the one-thousand-mile range limit. Three additional flights in October 1957 proved to be just as fickle as the five beforehand, with varying results. Missile nos. 107 (failure), 108 (partial success) and 109 (success) finished out the first phase of R&D (research and development) testing.[28] Listed below are the five series of tests and their aim during each stage, indicating an incremental approach to testing the missile subsystems:

Series I: The missile consisted of an airframe, propulsion system, radio guidance system and dummy warhead. Objectives were to evaluate performance of the airframe, propulsion and control system in flight and to evaluate the test procedures and ground support equipment.
(eight launches, two successes, one partial success and five failures for a 25 percent success rate)

Series II: The missile consisted of an airframe, propulsion system, all-inertial guidance system and dummy warhead. Objectives were to evaluate the all-inertial guidance system and reentry vehicle separation mechanism, as well as test procedures and ground support equipment.
(five launches, two successes, one partial success and two failures for a 40 percent success rate)

Series III: The missile consisted of an airframe, propulsion, all-inertial guidance system and functioning reentry vehicle. Objectives included determining performance of the reentry vehicle (including warhead) and demonstrating subsystem compatibility.
(five launches, four successes and one failure for an 80 percent success rate)

Series IV: The missile consisted of an operational system (with no warhead). The objective was to evaluate the performance of the entire weapon system to provide maximum confidence in system performance.
(twenty-eight launches, twenty-one successes, five partial successes and two failures for a 75 percent success rate)

Series V (R&D): The R&D missiles had a modified propulsion system (MB-3 Block II engine) and different G.E. nose-cone. The objectives included testing enhanced engine and evaluating reentry vehicle parameters.
(three launches, three successes for a 100 percent success rate)

Series V (Operational): Additionally, Series V included operational missile flights taken from Vandenberg AFB.
(twenty-two launches, eighteen successes, one partial success and two failures for an 87 percent success rate)

While on approach to Vandenberg AFB, a RAF Handley Page Victor flies over a readied Thor at Complex 75. *Courtesy of the United States Air Force.*

MISSILES AT VANDENBERG

Realizing that the facilities at Cape Canaveral were good for testing but not operational launches, the Air Force began a search in January 1956 for a more suitable location to launch operational Thor missiles.[29] Over two hundred government-owned sites were surveyed for favorable characteristics. Camp Cooke, a sleepy Army post on California's central coast, was selected due to its unique characteristics:

1) It was large and remote, at safe distances from nearby civilian communities (Santa Barbara, Lompoc, Santa Maria).
2) Adequate infrastructure—such as housing, barracks, railroad lines—were already present.
3) The post bordered the Pacific Ocean, an excellent testing ground for rockets and missiles.
4) The location was near aerospace manufacturers in San Diego and Los Angeles.

Additionally, by unique geographic quirk, the jut of land known as Purisima Point offered a launch site that could fire a rocket southward into a "polar orbit" without crossing any land mass until Antarctica. This specific orbit gave complete worldwide coverage to a space vehicle flying along it and originated an interesting offshoot of the space age: the advent of the reconnaissance (or "spy") satellite.

On November 16, 1956, SECDEF Wilson directed the northern portion of Camp Cooke be transferred to the USAF for missile development, specifically IRBMs and ICBMs.[30] The southern portion of the base was transferred to the U.S. Navy and redesignated Naval Missile Facility, Point Arguello (NMFPA).

Initial groundbreaking for new facilities took place on May 9, 1957. The installation was initially assigned to the Air Research and Development Command (ARDC), the major command overseeing the ballistic missile effort. In April 1957, ARDC established the 392nd Air Base Group at Cooke, followed three months later by the activation of the 704th Strategic Missile Wing and the 1st Missile Division. This buildup included a significant increase in the number of buildings, personnel and equipment on the base. Aboveground and underground missile launch complexes appeared as tons of concrete and steel shaped the landscape

On January 1, 1958, management responsibilities for Cooke AFB were transferred from ARDC to the Strategic Air Command (SAC). Along with the transfer, SAC acquired the three ARDC base organizations and responsibility for attaining initial operational capability (IOC) for the nascent U.S. nuclear missile force. In addition, SAC was directed to conduct training for missile launch crews, the men (and, later, women) in direct control of the missile complexes around the United States. Site activation tasks and research and development testing of ballistic missiles remained with ARDC. Space launches were to be conducted jointly by both commands. Although

What's in a Name?

From 1941 to 1946, and later from 1950 to 1953, the base was known as Camp Cooke, after Civil War Union general Philip St. George Cooke. After its selection and transfer by SECDEF Wilson for Air Force ballistic missile launches, the base was redesignated Cooke Air Force Base. In October 1958, the base took its present-day name of Vandenberg AFB in honor of General Hoyt S. Vandenberg, the second Air Force chief of staff. Missile trainees and on-base workers sometimes refer to it as Vandy-land, mirroring the name of a popular mouse-infested amusement park in Southern California.

the mission was now divided between the two commands, both SAC and ARDC maintained an integral relationship that was to blossom at Vandenberg over the next thirty-five years.

Construction at the newly renamed Cooke Air Force Base proceeded at a rabbit's pace, but arriving personnel had to make do without certain amentities. Housing, for example, was in short supply. Starting with eight homes on base, the number ballooned to six hundred within a year and a half, housing the families of the newly anointed "missilemen." General Mark Wade, commander of the base's 1st Missile Division, spoke about these New Age warriors at a dedication ceremony on October 5, 1958:

> *Here, the Air Force missileman will take his place beside the bomber crew and the fighter pilot who stand ready to retaliate against any aggression. In time to come, the missile badge he wears will become as well known as today's pilot wings, and* [this base] *will be known as the home of that new breed of airmen—the Missilemen....It will be a continuing challenge... to live up to the responsibilities...of a proud and priceless heritage and the beginning of a tradition.*[31]

Complex 75

As the Thor R&D program in Florida continued, the Air Force prepared for its eventual acceptance into the strategic deterrent force. In the SLC-10 Historic American Engineering Record (HAER), the initial mission of Cooke (later Vandenberg) AFB was "to produce an operational Thor weapon system and to train Thor IRBM combat crews from Great Britain's Royal Air Force (RAF)."[32]

To launch the Thor, contractors working under the U.S. Army Corps of Engineers (USACE) began work on the area known as Complex 75, after Thor's designation of SM-75. The pads were later renamed to the currently used term "space launch complex" in 1966. Complex 75 would contain seven launch pads: two each at 75-1 and 75-3 and three at 72-2 (SLC-2, SLC-1 and SLC-10, respectively).

On August 13, 1958, Thor vehicle No. 151 left Santa Monica, California, and traveled 160 miles north to Cooke Air Force Base. As one of the few bases in the USAF with a primary mission *besides* aircraft, the runway at that time was not prepared to host the large cargo aircraft used to airlift Thor

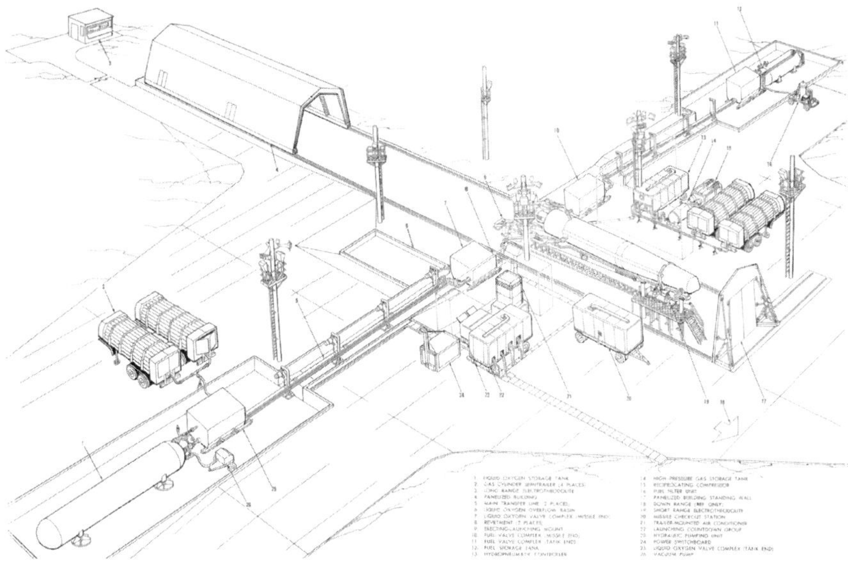

Thor Launch Emplacement and Ground Support Equipment

A technical diagram of a typical Thor IRBM launch site highlights the launch mount, moveable shelter and the two blast walls on either side of the launch pad. *Courtesy of Vandenberg Space and Missile Technology Center.*

hardware. (The base's runway would not officially open to air traffic until February 12, 1959.)[33]

Before the missile's arrival, missileers and maintainers were already training on using proper safety precautions for the deadly metal behemoths. Liquid oxygen–generating plants sprung up around the storage hangars, creating the oxidizers for the rocket engine's combustion chamber. Command destruct transmitters were located around Vandenberg; these high-powered "radio stations" could relay an electronic signal to a wayward missile and destroy it before it could have catastrophic consequences like impacting a populated area. Other safety concerns within Vandenberg's rapidly increasing missile complex were the safety of trains passing through the base along the Southern Pacific Railroad main line. Though Complex 75 resided west of the main line, any stray firings going eastward would be detonated immediately after launch, likely spraying metal fragments and rocket fuel over the railroad tracks.

First Thor Launch (N34.7516, W120.6193)

The first missile launched from Vandenberg (code-named "TUNE UP"), as well as being the first Thor IRBM launch on the West Coast, occurred on December 16, 1958, at Launch Pad 75-1-1 (later known as SLC-2E). This was also the first missile flight from Vandenberg. The repurposed launch site is approximately 1,600 feet southwest from the large blue-green gantry of SLC-2W.

IOC testing continued in Florida as the West Coast launchers were being built. Missiles No. 138 (November 5, 1958), 140 (November 26, 1958) and 145 (December 5, 1958) accumulated more data on the flight profile of the missile and warheads, with all being declared successful launches. A culmination point for the Thor program was reached on December 16, 1958. This day would find two Thor missiles prepared for launch, each a continent apart: one IOC missile at Patrick AFB (Thor 146) and one operational missile at Cooke AFB (Thor 151). While the IOC missile in Florida would be the first launched during the day, all eyes were on the West Coast. The first launch of an operational missile and first missile launch from an SAC crew on the West Coast was a success.

The operational testing phase at Vandenberg would see a total of twenty-two combat training launches, with eighteen successful launches, one partial success and two failures for an 87 percent success rate.[34]

Chapter 3

PROJECT EMILY AND THE FIRST MISSILEERS

Emily, to us, has been a gal of many moods, delightful at times and perfectly horrid on other occasions.
—W.L. Duval, Thor deployment manager

Strained Relations in the Special Relationship

During the early stages of the Thor program, U.S. war planners had to answer the question "Where should the missiles be located?" The plan for Thor's technolocal "cousins," the Atlas and Titan ICBMs, was already answered in the form of launch locations inside the continental United States. Their powerful engines and multistage designs provided enough thrust to loft the warheads a few thousand miles to their targets. With Thor's lesser range, it was imperative to place the launch locations closer to the targets. The most logical choice for basing close to the Soviet Union was either on the British Isles or within Europe proper. The Army's competing IRBM, the Jupiter, would later be placed in Turkey and Italy to strike Soviet targets. U.S. planners, as early as 1955, recognized the United Kingdom as a good location for Thor basing. Official discussions between the two governments did not take place until July 1956.

Indeed, the placement of Thor offered a chance to repair the political damages done to the U.S./British "special relationship," damaged during the 1956 Suez Crisis. As a declassified cryptological history of the Suez

Seven RAF members from No. 97 Squadron pose in front of Thor IRBM, serial number 214. The Combat Training Launch, code named PIPERS DELIGHT, lifted off from LE-8 on December 5, 1961. *Courtesy of Vandenberg Space and Missile Technology Center.*

Crisis from the National Security Agency (NSA) stated, the conflict caused a political dilemma: "The Suez crisis of 1956...presented the United States with a unique political and intelligence dilemma: two close U.S. allies, Britain and France, opposed American policy objectives. The battle for the Suez served as a model for examining the effect of political dissension and conflict on the intimate Anglo-American intelligence relationship."

The start of the crisis saw the nationalization of the Suez Canal by the Egyptian government, led by Gamal Abdel Nasser, who had orchestrated the overthrow of Egyptian King Farouk in 1954. Egyptian complaints about the occupation of the Suez Canal Zone saw the British remove their forces in June 1956.[35] In the lead-up to the crisis, the Egyptian government nationalized the Suez Canal Company, where the largest single shareholder was the British government. The British government saw this as a direct attack on its strategic interests in the Middle East, namely oil supply routes. Combined with the French government's disdain for Nasser and his assistance to the Algerian rebellion, the British planned a military intervention with French and Israeli assistance: Operation MUSKETEER.[36] The attack began on October 29, 1956, and lasted until a cease-fire was declared on November 6. In retrospect, the NSA study succinctly describes the quandary of allies and defense relationships:

> [W]*hether by dint of loyalty or the inertia inherent in any established system or bureaucracy, the Anglo-American SIGINT (signals intelligence) alliance was easily strong enough to continue unabated despite a disruption in the political relationship. In spite of British duplicity before and during the Suez crisis, our SIGINT interdependence was untouched by the temporary "spat" between allies. The interesting question raised by this phenomenon is, at what point is the* [defense] *relationship between allies affected by the current political environment?*[37]

The Bermuda Agreement

Changes in the post-Suez British government, namely Harold Macmillan becoming prime minister, offered an opportunity to mend the special relationship. Secretary of the Air Force Donald Quarles had unofficial briefings with the UK defense establishment as early as July 1956 about Thor basing in the United Kingdom, but the idea didn't take flight until 1957.[38] A meeting of Eisenhower and Macmillan on the island of Bermuda took place in late March 1957.[39] Meant as a public show of goodwill and strengthening relations between both countries after the Suez Crisis, the conference allowed the case of Thor deployment to be put forth and solidified. Author Boyes recounts the outcome of the Bermuda Conference: "The agreement made in Bermuda, nonetheless, was only an agreement in principle. It gave no indication on how the project would be managed or funded."[40]

An aerial view of a combat-ready Thor complex in the United Kingdom. Note the close distance between the three launch pads. *Courtesy of Vandenberg Space and Missile Technology Center.*

Specifically, the five-year agreement did not name Thor but stated that "in the interest of mutual defense and mutual economy certain guided missiles will be made available by the United States to British forces."[41]

Eleven months after the Bermuda Conference, Eisenhower codified the agreement with a signature in February 1958. The final result equaled sixty Thor IRBMs divided into four wings for use by RAF personnel. As originally envisioned by SAC, the Thor missiles would by operated by American personnel on U.S. airbases in the United Kingdom. As part of his administration's fiscal practicality and with the insistence of the United Kingdom, Eisenhower agreed that the United States would pay for all missile hardware (to include support equipment) and all training of launch crews and maintenance personnel. In turn, the United Kingdom would pay for base infrastructure construction and the costs of RAF personnel and civilian support positions.[42]

PROJECT EMILY

The operation to deploy Thor was called Project EMILY by the United Kingdom and Operation GO AWAY by the United States.[43] To move the materiel from California to England took the might of the Military Air Transport Service (MATS) and its fleet of C-124 Globemaster II and C-133 Cargomaster aircraft.[44] The airlift for Project EMILY required around 1,180 flights out of Long Beach Airport, located next to the Douglas Aircraft Company's facilities, to ease the transition from load-out to lift-off.

In addition to being the manufacturing hub for Thor equipment, Douglas Aircraft Company would send the first Americans to the United Kingdom for Project EMILY. They would take care of the matters needed to get the IRBM system to full operational capability before transfer to the RAF. American engineers and technicians provided critical continuity; as the ones who developed and manufactured the equipment, they would be the subject matter experts to answer any questions from the RAF.

A small exchange was located at every RAF base during Project EMILY to allow American personnel to purchase some of the comforts of home, such as tobacco and jewelry. *Courtesy of Vandenberg Space and Missile Technology Center.*

A typical Thor launch emplacement inside the United Kingdom during Project EMILY.
Courtesy of Vandenberg Space and Missile Technology Center.

Table 1. RAF Deployment of Thor IRBMs

Location	County	Squadron	Organized	Activated	Inactivated
Feltwell Group					
RAF Feltwell	Norfolk	77 (SM)	1-Sep-58	9-Jan-59	10-Jul-63
RAF Shepherds Grove	Suffolk	82 (SM)	22-Jul-59	22-Jul-59	10-Jul-63
RAF Tuddenham	Suffolk	107 (SM)	22-Jul-59	22-Jul-59	10-Jul-63
RAF Mepal	Cambridgeshire	113 (SM)	22-Jul-59	22-Jul-59	10-Jul-63
RAF North Pickenham	Norfolk	220 (SM)	22-Jul-59	22-Jul-59	10-Jul-63
Hemswell Group					
RAF Hemswell	Lincolnshire	97 (SM)	1-Dec-58	12-Jan-59	24-May-63
RAF Ludford Magna	Lincolnshire	104 (SM)	22-Jul-59	22-Jul-59	24-May-63
RAF Bardney	Lincolnshire	106 (SM)	22-Jul-59	22-Jul-59	24-May-63
RAF Coleby Grange	Lincolnshire	142 (SM)	22-Jul-59	22-Jul-59	24-May-63
RAF Caistor	Lincolnshire	269 (SM)	22-Jul-59	22-Jul-59	24-May-63

Location	County	Squadron	Organized	Activated	Inactivated
Driffield Group					
RAF Driffield	Yorkshire	98 (SM)	1-Aug-59	8-Jan-59	18-Apr-63
RAF Full Sutton	Yorkshire	102 (SM)	1-Aug-59	8-Jan-59	27-Apr-63
RAF Carnaby	Yorkshire	150 (SM)	1-Aug-59	8-Jan-59	9-Apr-63
RAF Catfoss	Yorkshire	226 (SM)	1-Aug-59	8-Jan-59	9-Mar-63
RAF Breighton	Yorkshire	240 (SM)	1-Aug-59	8-Jan-59	8-Jan-63
North Luffenham Group					
RAF North Luffenham	Rutland	144 (SM)	1-Dec-59	12-Jan-59	23-Aug-63
RAF Polebrook	Northamptonshire	130 (SM)	1-Dec-59	12-Jan-59	23-Aug-63
RAF Folkingham	Lincolnshire	223 (SM)	1-Dec-59	12-Jan-59	23-Aug-63
RAF Harrington	Northamptonshire	218 (SM)	1-Dec-59	12-Jan-59	23-Aug-63
RAF Melton Mowbray	Leicestershire	254 (SM)	1-Dec-59	12-Jan-59	23-Aug-63

In an interesting move, both governments allowed Douglas engineers to bring family members over to the United Kingdom for the project's duration. Approximately 1,200 personnel would travel to England, eventually living in modified trailer parks near the RAF Thor bases. Makeshift accomodations were created to support the personnel, such as medical trailers, in addition to engaging their social lives. Gathering spots were created, such as the Emily Club, run by elected employees, for recreation and relaxation.[45] A ladies' auxillary group ran daytime activities, such as tours of Britain and volunteer fashion shows. Mirroring similar circumstances during World War II, handfuls of eligible American bachelors arrived in the United Kingdom unencumbered and returned to the United States as married men.

Douglas engineers arrived on July 24, 1958, to begin the work required to get the sites built, readied and armed. The first Thor missiles in the United Kingdom arrived on August 29, 1958, at RAF Feltwell. C-124 Globemasters flew in from the United States to RAF Lakenheath, the base nearest to Feltwell that could accommodate the large aircraft and its Thor cargo. The first RAF missile unit, the 77th Strategic Missile Squadron, was formed at the base on September 1, 1958. Within the "spoke and hub" system, each deployment group had one central base and four outlying locations, each with three missiles. Four of these groups equaled the sixty missiles allotted to the RAF.

The agreement between the Eisenhower administration and the Macmillan government was amicable in many ways, but one area of concern was the "dual-key" functionality of the missiles. One part of the agreement was

SKY ZEST

VOL. 1 FEBRUARY 1959 NO. 1

PENTAGON OFFICIAL TOURS INSTALLATION

Donald Quarles, Deputy-Secretary of Defense, recently visited our project on a swing through overseas installations. Accompanying him on the local inspection are (left to right): Brig. Gen. R. H. Warren, Col. D. K. Lyster, Air Marshall Cross, Col. McDowell (partially hidden), W. L. Duval, Squadron Leader Baldock, J. H. Sogg, C. C. Martin, Quarles, Major General W. H. Blanchard, Commander 7th Air Division, T. S. Crispin and Col. W. A. Delahy.

NEWS FROM HOME

Through the efforts of the Management we are now receiving quantities of Los Angeles newspapers for distribution to as many employes as possible. The first shipment arrived on January 22 and was divided between the Caravan Camp, Lynford Hall, Brandon Park and the BOQ's.

Obviously, it would be impossible to get papers for everyone—so please pass them on to your neighbor and help us get the distribution as wide as possible. A dozen copies each of of "The Los Angeles Times" and "The Examiner" are arriving daily (flight weather permitting) and the employe interest in the home town news is in conversational evidence everywhere.

The first issue of the *SKYZEST* is part of our effort to let our fellow employes know "what is going on . . ." within our ranks. We hope to pass along the bits of information from time to time that have been so difficult to circulate because of the widespread activities of our operation.

Although we are not scheduling the first few issues of the *SKYZEST* to a date, we intend to put them out often enough to give you an interesting series to help you remember some of the highlights of your stay in England.

W. L. DUVAL,
Regional Operations Manager.

Left: *SKYZEST* was the Douglas Aircraft Company's "internal" newsletter for Project EMILY contractors and their families. *Courtesy of Vandenberg Space and Missile Technology Center.*

Below: The initial USAF cadre is briefed on the Thor system by Douglas Aircraft Company employees. Note the tip of the missile displaying a dummy shroud, not a Mark II reentry vehicle. *Courtesy of the United States Air Force.*

What's in a Name? (Part II)
John Boyes recounts from an interview with Rowland Hall, a member of the Air Ministry's Directorate of Works, that the name "Emily" had no hidden significance to the project. After being queried by a USAF lieutenant colonel for any project name ideas, Hall states, "I had been given a calendar with a scantily clad female on it, with the name of Emily in small print. I decided to stencil her name in large letters and [pin it] on the notice board."[46]

the United Kingdom would have control of (and the ability to) launch the missiles; however, the United States retained control over arming the 1.4-megaton W-49 nuclear warheads through an on-site authentication officer (AO). Before a launch, two keys would be turned: one by the RAF launch control officer to start the launch sequence and the other by the American AO to arm the nuclear warhead. This offered an interesting, as-yet-unheralded predicament: during a time of crisis elevation and possible nuclear conflict, either country could effectively negate the weapon system by refusing to "turn keys."

On February 11, 1960, the Air Ministry released the following statement: "The build-up of the [Thor] force now has

A Military Air Transport Service C-124A Globemaster II off-loads the sixtieth (and final) Thor IRBM in England for Project EMILY. *Courtesy of Vandenberg Space and Missile Technology Center.*

been completed. That this state has been reached within two years of the signing of the Inter-Government Agreement is a measure of the degree of cooperation between the United States Air Force and the Royal Air Force."[47]

392ND MISSILE TRAINING SQUADRON, THE FIRST MISSILEERS

While the missiles were arriving in England, RAF crews were being sent to California to begin training. The year 1957 was a whirlwind for many organizations on Vandenberg AFB. The USAF constituted (created a paper organization of) the 392nd Strategic Missile Squadron (Training) on May 23, 1957, during the buildup of facilities at then-Cooke AFB. The squadron was later redesignated as the 392nd Missile Training Squadron, obeying a Headquarters USAF decision that the term "Strategic" be used for organization names with direct operational employment of weapons. It is difficult to believe that it took almost four months before the organization activated officially (in September 1957) and eventually received its mission focus, operating the SM-75 Thor.[48] In an effort to accelerate the achievement of operational status following Sputnik, SAC had assumed responsibility for all ballistic missile training and deployment on January 1, 1958. That same day, SAC activated the 672nd Strategic Missile Squadron (THOR). The unit would later be renamed 672nd Technical Training Squadron and moved to RAF Feltwell to support USAF personnel in the United Kingdom until its inactivation in October 1959.[49]

A cadre of American officers and enlisted men attended instructor training at Sheppard AFB, Texas, in December 1957.[50] The RAF established an office on base around July 8, 1958, with Flying Officer Cyril D. Quinton earning the title as the first RAF liaison officer. Integrated Weapon System Training (IWST) for a combined USAF/RAF class began on August 20, 1958. Ten USAF and forty-seven RAF students (representing the RAF's 77th Strategic Missile Squadron) began training on procedures without utilizing actual hardware. The SLC-10 HAER report notes an interesting fact: the first IWST classes were ongoing before any of the launch facilities were constructed. Apparently this caused a bit of friction with the RAF personnel, especially their representative in Washington, D.C.[51]

The end of 1958 saw 245 RAF crew members completing training from the 392nd MTS cadre, with another large contingent still in training. While Thor IWST continued through the first quarter of 1959, a surprise occurred

A class picture of the 392nd Missile Training Squadron at Vandenberg AFB in September 1961. *Courtesy of Vandenberg Space and Missile Technology Center.*

The 392nd Missile Training Squadron Christmas party in 1961 shows happy children meeting Santa Claus. *Courtesy of Vandenberg Space and Missile Technology Center.*

392nd MTS students inspect hardware at a Thor launch site. *Courtesy of Vandenberg Space and Missile Technology Center.*

during a captive firing test. On April 9, Thor missile serial number 178 exploded at Launch Emplacement 2 (present-day SLC-2W). The accident destroyed the airframe; however, it did not slow down training or preparation for the first RAF launch, which was scheduled for the following week.

The RAF launches were divided into two types: IWST "training" launches and Combat Training Launches (CTLs). The IWST launches were performed by a class that was currently in training, whereas CTLs had "graduated" RAF missilemen come back to Vandenberg after being sent to the field. The IWST represented validation of the training, while CTLs validated the complete weapon system (human proficiency, hardware, procedures).

On April 16, 1959, Operation LIONS ROAR was successfully initiated by RAF crews at pad 75-2-8 (LE-8) as an IWST launch. Subsequent launches held at SLC-10 facilities included RIFLE SHOT on June 16, 1959 (at 75-2-7 [SLC-10E]), and SHORT SKIP on August 14, 1959 (at 75-2-6 [SLC-10W]). SHORT SKIP was unique, as it was the only RAF launch held at 75-2-6; all SLC-10 launches afterward were at 75-2-7 or 75-2-8.[52]

TABLE 2. RAF THOR MISSILE LAUNCHES AT SLC-10

Date	Nickname	Op. No.	Complex	Booster	Type
16-Apr-59	LIONS ROAR	1005	75-2-8	Thor	IWST
16-Jun-59	RIFLE SHOT	1008	75-2-7	Thor	IWST
3-Aug-59	BEAN BALL	1011	75-1-1	Thor	IWST
14-Aug-59	SHORT SKIP	1014	75-2-6	Thor	IWST
17-Sep-59	GREASE GUN	1016	75-1-2	Thor	IWST
6-Oct-59	FOREIGN TRAVEL	1018	75-2-8	Thor	CTL
21-Oct-59	STAND FAST	1019	75-1-1	Thor	IWST
12-Nov-59	BEACH BUGGY	1020	75-1-2	Thor	IWST
1-Dec-59	HARD RIGHT	1022	75-1-1	Thor	CTL
14-Dec-59	TALL GIRL	1023	75-1-2	Thor	IWST
21-Jan-60	RED CABOOSE	1025	75-1-2	Thor	IWST
2-Mar-60	CENTER BOARD	1028	75-2-8	Thor	CTL
22-Jun-60	CLAN CHATTAN	1034	75-2-7	Thor	CTL
11-Oct-60	LEFT RUDDER	1040	75-2-8	Thor	CTL
13-Dec-60	ACTON TOWN	1048	75-2-8	Thor	CTL
29-Mar-61	SHEPHERDS BUSH	1056	75-2-7	Thor	CTL
20-Jun-61	WHITE BISHOP	1063	75-2-7	Thor	CTL
6-Sep-61	SKYE BOAT	1072	LE-7	Thor	CTL
5-Dec-61	PIPERS DELIGHT	1087	LE-8	Thor	CTL
19-Mar-62	BLACK KNIFE	8201	LE-7	Thor	CTL
18-Jun-62	BLAZING CIDERS	9201	LE-8	Thor	CTL

The IWST training program ceased at Vandenberg on January 21, 1960, with the successful launch of the eleventh RAF Thor.[53] The final tally showed around 1,250 RAF missile personnel trained, along with "several hundred" American missilemen. While new crew training ended, RAF personnel still traveled to Vandenberg for combat training launches. As a precursor to the present-day force development evaluation launches (FDEs), the Thor program incorporated a number of live-fires for the weapon system. A Thor missile would be prepared and then raised to firing position, allowing the crew to work through their launch checklists, and then the missile would fly across the Pacific Ocean to a predetermined distance. Some of these tests would simulate combat firing conditions (to the maximum extent possible) and allow verification of both the missile and infrastructure's reliability and validation of the processes used to train the crewmen.

CTL missiles would be chosen from the deployed force and sent to Vandenberg for launch. The operation was an excellent continuation of skill validation for many areas—the crews to remove the missiles from alert status, the maintenance personnel to prepare the missile for shipping, the aircrews to transport the weapons back to the continental United States and Thor personnel at Vandenberg to receive and transport the missile to the launch pads. The cradle-to-grave care given in this manner presumably provided assistance in the repatriation of missiles after EMILY ended.

The twelfth CTL, code-named BLAZING CIDERS, took place on June 18, 1962, at pad 75-2-8. This would be the closeout of the RAF launch program (of a combined twenty-one IWST/CTL firings) and the last launch of a Thor IRBM at Vandenberg AFB.

END OF EMILY

While most of the Thor facilities were dismantled in the years following the end of Project EMILY, RAF Feltwell was home to many other space-related missions after Project EMILY. From 1989 to 2003, the USAF's 5th Space Surveillance Squadron (SSS) ran passive sensors for the Space Surveillance Network, monitoring satellites in orbit to track systems in both near- and far-Earth orbits. After the 5th SSS's inactivation, RAF Feltwell became home to Detachment 4, 18th Intelligence Squadron (IS), providing space intelligence information to U.S. Strategic Command and the National Security Agency. During Halloween, the giant radomes of the 18th IS are illuminated orange with a jack-o'-lantern's face.

In mid-1962, U.S. Secretary of Defense Robert McNamara informed the United Kingdom that support for Thor would end on October 31, 1964, ending the five-year agreement. According to Boyes, there seemed to be little support within the UK Ministry of Defence to take on the funding challenge for Thor, whose strategic obsolescence was becoming evident in the pace of technological advances.

As the American ICBM force grew in number and survivability, the stop-gap nature of the Thor deployment to the United Kingdom wore thin. Following on the heels of Thor's acceptance into the nuclear force, the Atlas-D ICBM was declared operational by Commander-in-Chief, Strategic Air Command (CINCSAC), General Thomas S. Power after a successful launch at Vandenberg AFB on September 9, 1959.[54] The Titan I ICBM was declared operational on

Above: A C-124 Globemaster II cargo aircraft delivers a Thor missile to the United Kingdom during Project EMILY. *Courtesy of Vandenberg Space and Missile Technology Center.*

Right: The cover of the Project EMILY "yearbook" created for the Douglas Aircraft Company members and their families. The book was instrumental in recording the experience of U.S. workers in the United Kingdom. *Courtesy of Vandenberg Space and Missile Technology Center.*

April 18, 1962.[55] These two first-generation ICBMs would not last long in the operational force. Both would be withdrawn completely by 1965, in favor of the silo-launched second-generation missiles, the Titan II and Minuteman. Finally, increasing mission requirements created a need for additional Thor boosters. Instead of programming additional airframes into the budget, reusing the EMILY boosters seemed the wisest course of action for quick reaction programs like DMSP, Program 437 and the Pacific nuclear tests.

Inside the final edition of the monthly Project EMILY newsletter, W.L. Duval, Thor deployment manager for Douglas, wrote a bittersweet soliloquy for "her":

> *A feminine personal name, according to Funk & Wagnall, but to those of us who have known EMILY intimately, this is a pretty inadequate definition. Emily, to us, has been a gal of many moods, delightful at times and perfectly horrid on other occasions. She was a thing of beauty, an ugly wench, a lovely lady, a nagging shrew, a light-hearted child, a swearing, worrying mother. She could be warm as a sunny summer's day, cold as a midwinter night, tender as the gentlest lover or harsh as the toughest gun-moll. Yet, in spite of her many faceted personality, I, for one, loved her…as I'm sure many of you did. She was, to say the least, an interesting female, and one we shall never forget. I am sad that the time has come to leave her. Many thanks to all of you for your help in taking care of her during our whirlwind courtship.*[56]

Chapter 4
NUCLEAR PLAYGROUND

Caveat Aggressor ("Aggressors Beware")
—motto of the 10th Aerospace Defense Squadron

The launch of Sputnik on October 4, 1957, spread fear throughout the U.S. defense establishment—first because of the Soviet Union's ability to launch a payload into orbit, secondly due to the range of the rocket (modified ICBM, R-7 Semyorka) and finally because of the size of the proposed payload capability. While the satellite itself did little more than beep, the thrust provided by the R-7's numerous engines proved that the Soviets could launch heavy payloads into space. One idea postulated was that the Soviets were developing an orbital nuclear bombardment system for placing a nuclear warhead in orbit to be used at the time and place of their choosing. While the mathematic calculations of such a venture were complex, the simplest answer of "Yes, it's possible" was demonstrated in late 1957 by a small metal sphere flying around Earth every 96.2 minutes. The threat of orbital weapons went from theoretical to practical in a short amount of time.

Operation DOMINIC

While nuclear testing in the late 1950s and early 1960s focused primarily on weapon characteristics and delivery vehicles, a small subset of tests pondered the effect (and practical usage) of orbital nuclear explosions.

A Douglas Aircraft Company launch crew poses in front of the Johnston Island Thor launch emplacement during Operation DOMINIC in 1962. *Courtesy of Vandenberg Space and Missile Technology Center.*

Each test program detonation or "shot" initiated by the Atomic Energy Commission (AEC) relied on new designs, employment or delivery, enabling a massing of information on the use of and survival from atomic weapons. From November 1958 to September 1961, a voluntary test moratorium gave the United States and the Soviet Union a chance to create new weapon designs. One drawback of the thirty-four-month hiatus was the inclusion of new warheads into the stockpile that were not verified by a full-scale nuclear test. Designs were incomplete, and some contained errors that prevented the weapon from developing its full yield.

Quoted from the Defense Nuclear Agency's history of Operation DOMINIC:

> *On November 1, 1958 the United States began a unilateral moratorium on nuclear tests. The moratorium was linked to the issues of nuclear disarmament and the political struggle between the United States and the U.S.S.R. as well as concerns over the increasing levels of worldwide radioactive fallout.... On January 3, 1960, the Soviet Premier pledged*

that the Soviet Union would not resume testing unless the Western nations started first. In February, the U.S. President proposed a treaty banning all atmospheric tests and those underground tests powerful enough to register above 4.75 on the Richter earthquake scale.[57]

The Soviet Union broke the moratorium first. The United States followed with a series of thirty-one underground detonations under Operation Nougat at the Nevada Test Site (and one in a cavern near Carlsbad, New Mexico).[58] During the moratorium, another atmospheric test program code-named DOMINIC was planned in secret. On August 31, 1961, the Soviet Union announced it was resuming atmospheric testing. The United States had not expected the announcement, as its testing support organizations and infrastructure had deteriorated badly in the intervening years. Operation DOMINIC suffered from this sense of urgency, allowing efforts to be made on a "crash" basis with less than optimal planning, quick decisions and shortcuts and tighter-than-normal security. An early 1962 schedule was created for DOMINIC, titled "A Plan for Establishing a Capability for Large Yield Testing in Two Months," authored by the Sandia Corporation.[59]

Joint Task Force 8 (JTF 8) was established by the joint chiefs of staff to administer the DOMINIC series. JTF 8 consisted of military personnel from all four services, contractors and civilians from the DOD, the AEC, the U.S. Public Health Service and contractor organizations.[60] DOMINIC involved thirty-six nuclear test shots in the Pacific Ocean to include high-altitude tests and airdrops. Inside the plan, a capability for "out-of-the-atmosphere" (exo-atmospheric) testing was designed to "reduce fallout in the event surface testing is not authorized." A "permanent" Thor launch vehicle capability at Johnston Island was called out within the document:

Alternative Test Programs—(1) Short time scale proof tests of weapons systems, including a Polaris, Nike Zeus, and Minuteman at JI; (2) Thor with modified re-entry vehicle payload using mobile launch/ground support equipment, ground-based diagnostics, and companion rockets for measuring yield and gross case performance, could be conducted at JI within six months; (3) Thor missiles with specially-designed final stage vehicle (for testing large yield devices) using mobile launch/ground support equipment, could be conducted at JI within 12 months.

Payload Capability of Thor—Thor could provide the United States with an out-of-the-atmosphere testing capability with considerable growth potential in terms of device weights. It is estimated that the mobile launch/

The view of Johnston Island on a northeast approach shows the Thor launch emplacement in the far left of the photograph. *Courtesy of Vandenberg Space and Missile Technology Center.*

> *ground installation could be made permanent in 6 to 10 months at a cost of $10 million.*[61]

WHY JI?

Before its use in nuclear testing and for Program 437, JI's location in the Pacific Ocean was known to the western world as early as 1598, when Dutch sailors ventured past the atoll while searching for a route to Japan:

> *Captain Jacob Mahu sailed from Rotterdam with three ships to attempt a passage of the Pacific to Japan. One vessel was seized by the Portuguese in the Moluccas. By that time Mahu had died of disease, and command of the surviving vessels passed to a Captain Huydecoper. On November 27, 1599, these ships departed an island off the Chilean coast on a direct course for Japan. Several months later the* Hope *and* Charity *"fell in*

with certain islands in 16 degrees north, the inhabitants of which were man-eaters." Huydecoper shied away from contact with the cannibals, but eight sailors deserted from one of the ships in a pinnace and fled to the atoll, where they presumably fell victims to the inhabitants.[62]

Once the expedition's sole remaining vessel, *Hope*, reached Japan in the spring of 1590, the location of the atoll was made known via two letters from the chief pilot, William Adams. The islands, however, would exist in relative obscurity for the next two centuries. The British navy takes credit for officially discovering JI on December 14, 1807. Captain Charles Johnston, of the *Cornwallis*, described two islands at 16 degrees, 53 minutes north, 169 degrees, 31 minutes west. The pair of islands was dubbed Cornwallis and Johnston Islands after the ship and captain, respectively.

During the mid-1800s, the island received American visitors. According to Robert Streeter's "Johnston Memories" website:

On March 19, 1858, the captain of the American schooner Palestine *took possession of the islands in the name of the United States. Three months later, June 14 to 19, 1858, the Hawaiian schooner* Kalama, *Captain Watson, with Samuel C. Allen on board, visited Johnston, removed the American flag, and hoisted that of Hawaii. The larger island was renamed Kalama Island, and the nearby smaller island was called Cornwallis.*[63]

Coming back to the atoll in July 1858, the *Palestine* reaffirmed the gaining of the island for the United States and hoisted the colors. The same day, King Kamehameha IV annexed the island for the Kingdom of Hawaii, claiming it was derelict. While swapping national flags seemed like an irrelevant point to a tiny island in the Pacific, several centuries of large bird populations living on the island left tons of guano, an excellent fertilizer. Phosphates and ammonium compounds were also used as key ingredients in the manufacture of explosives, such as gunpowder. The material was highly sought; in 1856, the United States Congress passed the Guano Act.[64] The act authorized American citizens to claim unoccupied islands in the name of the United States for the purpose of obtaining guano:

Forty-eight Pacific islands were eventually claimed under the auspices of the Guano Act. William Parker's San Francisco–based Pacific Guano Company was organized to dominate the trade, and in 1859 the firm's first substantial shipment of guano (fifty tons) was extracted from Johnston

> *Island. Parker had competition from King Kamehameha IV of Hawaii, who also claimed possession of the atoll. The issue remained in dispute until 1898, when American annexation of Hawaii confirmed the company's claim to the guano beds.*[65]

While JI had infrequent visitation throughout the 1800s, it wasn't until the twentieth century that the atoll was noticed by global powers. In the 1930s, Pan American Airways flew routes from San Francisco to Hong Kong, using Hawaii as a refueling stop. Johnston Island was bypassed in favor of using either Midway or Wake Islands. The U.S. Navy, wanting to capitalize on the strategic positioning of the atoll, sent four seaplane squadrons to survey the islands as a forward airfield, in lieu of the Japanese Empire's regional ambitions. American naval strategists surmised that JI would be an ideal location for advance aerial patrols against the Japanese navy.

Not until March 1941 did Rear Admiral P.N.L. Bellinger, commander, Naval Base Air Defense Force, Pearl Harbor, continue the program of improvements on the island. Bellinger anticipated a possible strike against Pearl Harbor by Japanese aircraft carriers in the event of war and saw Johnston Island as an ideal base for staging aerial patrols against such a threat.

While JI played no direct part in the battle in the Pacific, the location remained critical as a temporary refueling stop for patrol and transport aircraft. In the spring of 1943, small detachments of PV-1 Ventura bombers were posted on JI to conduct antisubmarine patrols. After World War II, the atoll fell out of favor again. Administration of it was transferred to the Air Force in mid-1948, with aims to reduce the island to "caretaker" status. Changes in world politics saw JI become an important center to the HARDTACK series of nuclear tests.

Prior to DOMINIC, the HARDTACK test series had been successfully conducted from the island. During HARDTACK, two high-altitude bursts in the megaton range were set off in the vicinity of the atoll. The first device, code-named TEAK, was detonated on August 1, 1958, at 252,000 feet above the Pacific. The second shot, designated ORANGE, was exploded at 141,000 feet on August 12, 1958.[66] JI would host high-altitude tests with rockets and missiles, while the Christmas Island, a British territory, would host the air and naval weapon development tests.[67] According to the *New York Times*, JI was chosen due to its austerity and lack of indigenous population:

> *JI had been used in 1958 for rocket-borne, high-altitude nuclear tests and was available for this kind of testing for DOMINIC, but* [airborne and naval] *weapon development shots posed a more difficult problem. These could not be conducted off JI because they would interfere with the extensive preparations required for the high-altitude shots. The area that had been developed and used for large atmospheric tests, the atolls of Eniwetok and Bikini in the Marshall Islands, was no longer available. Eniwetok had been kept in a ready-to-test condition until 1960 when it had been turned over to the Air Force for its missile development programs. This area could have been recovered from the Air Force, but the United States was quite sensitive to the fact that the Marshalls were held by the United States as a Trust Territory, and the reintroduction of nuclear tests into the area would expose the United States to a great deal of unfavorable criticism.*[68]

With the importance of the upcoming test series, JI grew accordingly in size and population. According to Dr. Austerman's report of JI's history:

> *By early 1962, JTF-8 had taken charge of the atoll and a major facilities upgrade was in progress to prepare it as the site headquarters for a new round of tests. Johnston Island was soon inundated with personnel and equipment as a massive construction effort changed the face of the atoll. Very soon the spit of land was better than doubled in size as coral dredging and crushing added new area to the island. What had originally been a 500-foot runway eventually spanned 10,000 feet. The population was measured in the hundreds as billets, offices, warehouses, and workshops joined the new launch control facilities that sprouted up among the coarse grass and sandy hummocks. All work revolved around the launching pads that were built on the island's western side to receive the Thor IRBMs that were used to loft the test weapons to altitudes of several hundred miles for detonation.*[69]

Life inside a FISHBOWL

Tests would include two high-altitude detonations aboard Thor missiles, three surface blasts from Navy ships, one deep underwater shot and a large number of airdrops and balloon-borne shots. Four high-altitude/exo-atmospheric tests were collectively called FISHBOWL and used four

Thor boosters from a modified launch site. The Air Force dismantled the facilities from pad 75-2-6 (SLC-10W) and moved them to JI in early 1962 in preparation for DOMINIC. The pad was given the designation Launch Emplacement 1 (LE-1). As these prepatory moves continued for the DOMINIC series, the two remaining pads at Vandenberg (SLC-10E and LE-8) continued RAF training.

During the launch activities, around eight hundred nonessential personnel were taken off JI and moved to a nearby Navy ship. For the initial launches, personnel were transferred to the USS *Iwo Jima*. After BLUEGILL PRIME, the USS *Princeton* covered evacuation duties until the end of the DOMINIC series.[70]

One Thor "practice" shot, sans the nuclear warhead, took place on May 2, 1962. Code-named TIGERFISH, the Thor carried an inert warhead and instrument pods through the full launch sequence. The prime reason for the test was checking the pods against the surface and aerial instrumentation array, a gathering of ships and planes with recording equipment aboard.[71] These pods would be jettisoned from the missile before the nuclear detonation. While floating on parachutes, they would record nuclear characteristics, such as neutron flux and gamma radiation, from the blast and be recovered by the U.S. Navy.

BLUEGILL

Just after midnight on June 2, 1962, the first FISHBOWL test, designated BLUEGILL, ended in dramatic fashion when the range safety officer detonated the safety explosive package.[72] Since the instrumentation array of ships and planes were deployed, the JTF commander ordered the destruction of the missile (and warhead). One bright spot from BLUEGILL was the instrumentation pod recovery—all three survived and were brought back to JI.

STARFISH AND STARFISH PRIME

The next launch, STARFISH, took place on June 19, 1962. Two of the three instrumentation pods were replaced with experimental reentry vehicles. The launch occurred just before midnight local time, and the Thor

A (Nuclear Test) by Any Other Name…

The BLUEGILL test had a very unique naming convention, one that may not be clearly understood outside the scientific and engineering communities. The second attempt was named BLUEGILL PRIME, while BLUEGILL DOUBLE PRIME was attempt number three. Finally, the successful launch was under BLUEGILL TRIPLE PRIME. To recap: it took ***four*** Thor launch attempts, and three destructed warheads, to perform a single successful nuclear detonation for BLUEGILL.

A nuclear-laden Thor IRBM sits on Launch Emplacement 1 at Johnston Island before the ill-fated STARFISH shot on June 20, 1962. *Courtesy of Vandenberg Space and Missile Technology Center.*

held steady for almost a minute before the engine stopped and the airframe disintegrated in mid-flight. Ironically, the range safety officer destroyed the missile to prevent any further damage; however, plutonium-contaminated debris from the flight littered the island and surrounding ocean area.[73]

The next flight, code-named STARFISH PRIME, made its way into the history books after its launch on July 8, 1962. This was the first successful high-altitude detonation of the DOMINIC series. The blast of the 1.4-megaton warhead at its 250-mile altitude set off an electromagnetic pulse (EMP), a burst of electromagnetic energy that disrupted communications in the Pacific and destroyed unshielded electronic equipment. The EMP fried over three hundred streetlights in Hawaii (800 miles from ground zero) and wreaked havoc on the electrical grid.[74] The test also directly bombarded nearby orbital satellites with high-energy electrons, severely lessening their projected lifespans. Magnetic fields that surround Earth became highly charged with energetic particles after the test, causing long-term effects against satellites with little shielding against such an occurrence. The visual display caused widespread auroras and changed "night into day" for nearly six minutes, as recounted by observers in Honolulu.[75]

Additional effects discovered from the resultant nuclear explosion and subsequent EMP remained classified but were carefully added to the consideration of adopting nuclear-based weapons to damage space systems. The results of the STARFISH PRIME test led directly to the development of Program 437 as an antisatellite (ASAT) system.

BLUEGILL PRIME, DOUBLE PRIME AND TRIPLE PRIME

The next FISHBOWL shot, designated BLUEGILL PRIME, was attempted on July 25, 1962. This launch repeated the profile and test from the earlier June 3 attempt and incurred a similar fate. The Thor missile failed to provide sufficient thrust to lift off from the pad. Instead of shutting down the engines remotely, the range safety officer incorrectly forced the missile to self-destruct right atop LE-1. Radioactive particles were spread far and wide, and the launch emplacement took severe damage. JTF-8 planning had taken into account the protection of personnel by providing radiation-safe (RADSAFE) clothing during DOMINIC. The clothing was made of cloth/plastic and had tight wraps around the wrists, ankles and neck and included respirators. Contaminated outfits were either decontaminated or completely discarded as radioactive waste. Cleanup after BLUEGILL PRIME gave the RADSAFE clothing a workout. In a location where the military uniform consisted of short-sleeve shirts and shorts, tropical conditions made work uncomfortable while in the RADSAFE garb. After the explosion, radiation contaminated the area; cleanup consisted of soil removal, scrubbing (and later painting) contaminated surfaces and soaking soil with oil. In the aftermath, the two remaining launch pads from SLC-10 were dismantled and transported to JI to allow the program to continue. However, two months would pass before the next FISHBOWL shot.

What Were They Thinking?

The nuclear tests between BLUEGILL PRIME and KINGFISH took place during a dangerous time in world history: the Cuban Missile Crisis. During October 1962, American, Soviet and Cuban leadership faced off militarily and politically regarding secret nuclear missile deployments on the island of Cuba. DOMINIC testing continued during the crisis, however, with seven aboveground nuclear detonations; the Soviets also continued with their developmental testing.

BLUEGILL DOUBLE PRIME was scheduled for October 15, 1962. Continuing the spate of bad luck, the Thor booster was destroyed 156 seconds after lift-off. Radioactive debris from the destroyed rocket fell back on JI, requiring yet another testing pause for decontamination of the launch site.

The final test and last Thor launch, BLUEGILL TRIPLE PRIME, took place on October 26, 1962. This was the fourth attempt of the BLUEGILL series and the only successful one. Detonation occurred at an altitude of roughly thirty-one miles above the ocean surface and nineteen miles southwest of JI. The W-50 warhead released a yield in the "sub megaton" range.

AN EMPTY FISHBOWL

The next successful Thor launch took place on November 1, 1962, code-named KINGFISH. In conjunction with the launch, twenty-nine small rockets were launched to perform atmospheric and cloud sampling in the vicinity of the detonation. While the FISHBOWL series continued for one more shot (TIGHTROPE aboard a Nike-Hercules missile), operations involving Thor ceased. The launch emplacements used for DOMINIC remained on JI for undetermined future purpose. Recorded in the history books, DOMINIC was the last atmospheric nuclear testing conducted by the United States and by far the largest. The Limited Test Ban Treaty was signed the following year by the United States and Soviet Union, halting further atmospheric testing of nuclear weapons. Data derived from the FISHBOWL tests, however, allowed the United States to consider Thor for another mission: Program 437.

PROGRAM 437: THE USAF'S FIRST ANTISATELLITE WEAPON

During the earliest days of missile development, a heady truism was discovered: it was *very* difficult to shoot down incoming missiles. The speed of reentry made targeting and interception nearly impossible by non-nuclear means. Development of antiballistic missile (ABM) systems occurred almost

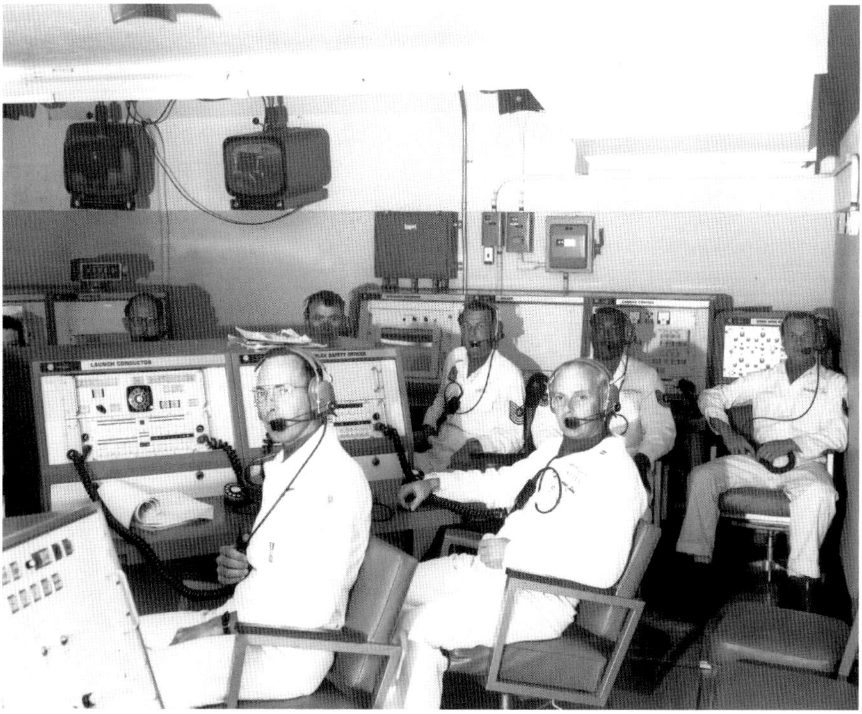

Program 437 crews take a break for a photo opportunity while training inside the East Blockhouse. *Courtesy of Vandenberg Space and Missile Technology Center.*

concurrent with the development of the missiles themselves, as protection against enemy systems. When the first satellites were launched into orbit, it was feared by American leadership that the Soviet Union would place nuclear weapons in orbit to rain them down at the time and place of their choosing. Creation of an ASAT weapon became a national imperative to allay fears that there was no defense against these orbital nukes.

In the early 1960s, SECDEF Robert McNamara gave approval for the U.S. Army to test ABM missiles, such as the Nike Zeus, as antisatellite weapons. Designated Program 505 and Project MUDFLAP, the Nike-Zeus missiles had an altitude range limitation of two hundred miles.

Saber-rattling by the Soviet Union on satellite destruction saw its Air Defense Forces (PVO Strany) stand up a new division in early 1963, "which was to be equipped with special spaceships, satellite fighters and other flying apparatus armed with rockets and radio-electronic apparatus."[76]

In September 1962, Air Force leaders submitted a proposal for using nuclear-tipped Thor missiles as ASATs. While the idea of using nuclear warheads to "kill" satellites was not popular, stemming from the STARFISH PRIME test results, Secretary of the Air Force Eugene Zuckert requested a plan.

In October 1962, the world watched as a nuclear crisis unfolded around the tiny island nation of Cuba. The seemingly irrational plan to secretly introduce nuclear-tipped IRBMs to the island showed the world that the Soviet Union would not take the idea of orbital nuclear weapons off the table. By February 1963, Program 437 had been presented to the Air Force chief of staff: "[D]evelopment of an operational capability to negate satellites has top priority among defense programs."[77]

The Kennedy administration, reeling from the figurative fallout from the Cuban Missile Crisis, was hesitant to have an antisatellite weapon system, let alone a nuclear one operated by "blue suiters" on alert. The discussion between administration officials was interrupted by U.S. Information Agency director Edward R. Murrow: "If the Soviets place a bomb in orbit and threaten us and if this administration has refused to develop a capability to destroy it in orbit, you will see the first impeachment proceeding of an American President since Andrew Johnson."[78] Program 437 was subsequently approved by SECDEF McNamara and the cabal of administration officials present.[79]

Officially, Program 437's primary mission was to destroy foreign-launched satellites when directed by commander in chief, Continental Air Defense Command (CINCONAD), by means of a non-orbiting nuclear warhead payload. The operational Program 437 missiles were located on JI (N16°43'46.4", W169°31'58.4") in the Pacific Ocean. The weapon system consisted of two Thor LV-2D launch vehicles, associated shelter and ground equipment and launch control facilities. Additional boosters were located at Vandenberg as operational ready back-up and for training at the 10th Aerospace Defense Squadron (ADS) facilities at SLC-10. On JI, two Thor missiles were used in a dual configuration with a "hot" back-up; this meant that both missiles were prepared for launch during an event as insurance that if one failed during the launch sequence, the other would successfully take off and intercept the hostile satellite. The initial reaction time from alert order (ALERTORD) to launch was estimated at two weeks (!), but further refinement took that down to no more than seventy-two hours from warning to weapon system deployment from Vandenberg storage.

In late April 1963, Air Force Systems Command (AFSC) activated the 6595th Test Squadron at Vandenberg to make Program 437 operational. The unit was originally authorized 18 officers and 135 airmen; many

The Thor emplacement at Johnston Island for Program 437. Note the short distance from the launch pad to the personnel barracks in the background. *Courtesy of Vandenberg Space and Missile Technology Center.*

were veterans of the Thor launch program and the BOMARC surface-to-air missile system. To complement the AFSC unit, SAC activated the 4300th Support Squadron at Pad 7 (aka 75-2-7 or SLC-10E) to assist with maintenance and launch training. A modified warhead was provided by the Sandia Corporation via the AEC.

As initially planned, the majority of the manning would remain at Vandenberg for launch training while a small contingent stayed at JI to provide caretaker tasks to the launch sites. Four missiles were originally projected for Program 437; however, with the cessation of Project EMILY in 1963, sixteen more missiles would be available for testing and operations. Projections for the first combat-ready launch were aimed at April 1964, with IOC obtained in May 1964 and the system fully operational by October. SECDEF McNamara expressed displeasure with the aforementioned plans, desiring a capability to "initiate destruction by a telephone call."[80] Employment timelines were slashed to four to five hours but no later than twelve hours from initial ALERTORD. Pushing back on McNamara's demands, analytic personnel stated that thirty-six hours of radar tracking and calculations were required for accurate orbital predictions.

In August 1963, a Program 437 "Phase II" plan was released by AFSC. The new plan had the operations squadron stationed permanently at Vandenberg, with launch crews and maintenance personnel rotating to JI on a ninety-day basis. The total personnel within the squadron equaled 468, with 79 "permanent party" on JI. From the remaining 389 personnel in California, over a third would be on JI on temporary duty status (TDY).

At Vandenberg, one Thor booster and its associated ground equipment were kept in a training simulation status, so no targeting computer was provided, nor was the "dual configuration" training performed. Ground guidance and destruct telemetering was simulated by the training equipment console, and the Ground Guidance Station sent signals to the missile to simulate in-flight steering actions. While at Vandenberg, personnel would train on control center and timing equipment functionally identical to the launch equipment at JI. The site setup duplicated the operational site as near as possible. Differences did exist, however, due to the unique requirements of the SLC-10 facilities as a dual training/launch location. One "minor" difference from the JI facilities was in the layout of the control center and direction center inside the Blockhouse. While on JI, Aerospace Defense Command's Space Detection and Tracking System (SPADATS) provided the orbital "shooting solution" for the hostile satellites. There was no such live-feed for 437 trainees at SLC-10 to work from.

Once Phase II was progressing successfully, changes at Vandenberg reworked organizational structures. The 6595th Test Squadron was inactivated and replaced by the 10th Aerospace Defense Squadron. The unit was the only "blue-suit" (composed entirely of military personnel) space launch organization in the entire Air Force. The mission's critical nature to national security gave impetus to a hectic pace of operations: "The squadron members joked that in their case the acronym 'ADS' meant 'All Day Saturday and All Day Sunday' as the tempo of training accelerated in 1964."[81]

The first demonstration launch was planned for Valentine's Day, February 14, 1964. The tests were given the strange-sounding code name of SQUANTO TERROR[82] and conducted over JI so the island's Baker-Nunn optical tracking camera could provide positive photographic proof of its success (or failure). Two missiles were in concurrent countdowns as launch time approached. The primary missile launched successfully and met an interception point at an altitude of 540 nautical miles above the Earth.

Personnel stationed on Johnston Island partake in a softball game. Off-duty personnel had their selection of many recreational activities while stationed in the Pacific. *Courtesy of the United States Air Force.*

The simulated warhead passed close enough to its target, a discarded rocket body, to be considered a success had a nuclear warhead detonated. The second test, occurring on March 1, 1964, had the primary missile taken offline due to instrument malfunction but allowed the second booster to meet the interception point. Cut-off of the main engine 4.61 seconds early gave a slight cross range error, but in the era of nuclear weapons, the Spherical Error Probable of 3.4 miles was deemed acceptable. The third SQUANTO TERROR test took place on April 23, 1964, with an all blue-suit crew from the 10th ADS. The launch and interception was successful.

The last test launch was scheduled for May 28, 1964, with a distinguished visitor in attendance: Lieutenant General Herbert Thatcher, the commander of Air Defense Command. The launch of the Thor vehicle failed; however, Lieutenant General Thatcher reviewed the post-launch evaluation. He determined that the failure was not on the crews or procedures but on the hardware; with this, he declared Program 437 had obtained IOC. Two weeks later, a second Thor missile was transported to JI to maintain a two-sortie, twenty-four-hour alert schedule.

Life on JI

When the launch crews were not preparing for test launches or running countless drills of procedures and checklists, there was time for activities on JI. According to Eric Lemmon, 10th ADS launch crews would rotation out to JI at ninety-day intervals:

> *The standard practice during the lifetime of Program 437 was to have three combat crews, each consisting of about 125 people in A, B and C Crews. If A crew was on JI, then B Crew would be at SLC-10 West launching DMSP spacecraft on a slightly different version of Thor, and C Crew would be at SLC-10 East getting trained and certified to replace A Crew on JI.*[83]

While at JI, there were plenty of physical activities to do when personnel were not on crew. Its position in the Pacific Ocean came with the (dis) advantages of being located on an island that was just shy of six hundred acres in size. Water sports (skin diving, scuba diving, fishing and boating) were extremely popular. The Pacific Atoll Divers Club provided equipment to interested divers, and diving practice was held in the outdoor pool before braving the ocean. Some of the more interesting hazards around JI included sharks, barracuda, moray eels, jellyfish, cone shells and coral. One humorous entry in the Johnston Atoll Safety Orientation guide: "Considering the number of people swimming in shark-infested waters and the number of actual shark attacks each year, your chance of getting bitten by a shark is the same as contracting a social disease on Johnston Island."[84]

Even with the mission, location and variety of activities, some personnel did not take to duty on JI very well, according to Lemmon:

> *Some crew members hated the ninety-day tours at JI, and complained bitterly about the separation from family and the boredom. A few had nervous breakdowns, and a few others had domestic issues to deal with. The majority of the* [personnel] *on the crew took the JI rotations in stride and were outstanding performers who took their jobs seriously. It takes a special breed of cat to be comfortable with weapon systems.*

A group of spirited individuals from Thor squadrons took special time to mention JI's wonderful characteristics in a "deed" given to the incoming combat crews:

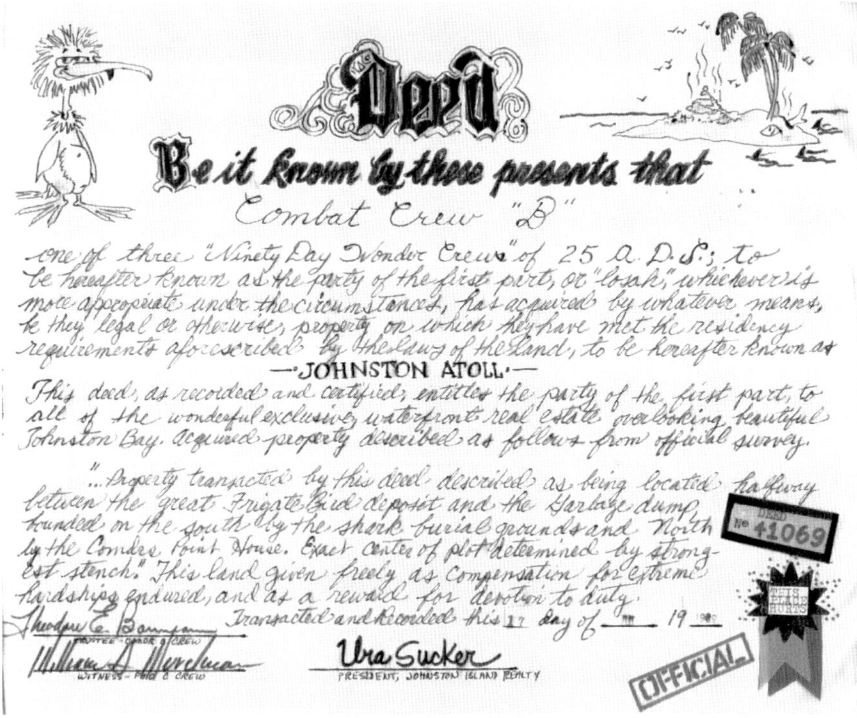

Deed

Be it known by these presents that

Combat Crew "D"

one of three "Ninety Day Wonder Crews" of 25 A.D.S.; to be hereafter known as the party of the first part, or "losah," whichever is more appropriate under the circumstances, has acquired by whatever means, be they legal or otherwise, property on which they have met the residency requirements aforescribed by the laws of the land, to be hereafter known as

—JOHNSTON ATOLL—

This deed, as recorded and certified, entitles the party of the first part, to all of the wonderful exclusive, waterfront real estate overlooking beautiful Johnston Bay. Acquired property described as follows from official survey.

"...Property transacted by this deed described as being located halfway between the great Frigate Bird deposit and the Garbage dump, bounded on the south by the shark burial grounds and North by the Comdrs Point House. Exact center of plot determined by strongest stench." This land given freely as compensation for extreme hardships endured, and as a reward for devotion to duty.

Transacted and Recorded this 17 day of ___ 19__

Ura Sucker

PRESIDENT, JOHNSTON ISLAND REALTY

DEED No 41069

OFFICIAL

This comical "deed" to Johnston Island was passed between combat crews by those leaving the atoll. *Courtesy of Vandenberg Space and Missile Technology Center.*

> *Property transacted by this deed described as being located halfway between the great Frigate Bird deposits and the garbage dump bounded on the south by the shark burial grounds and North by the Commander's Point House. Exact center of plot determined by strongest stench. This land given freely as compensation for extreme hardships endured, and as a reward for devotion to duty.*[85]

FROM SATELLITE INTERCEPTOR TO SATELLITE INSPECTOR

The Program 437 nuclear antisatellite mission lasted until 1965, when it began to have resources siphoned into offshoot projects. While a lack of Thor boosters for the program was known, additional procurement of boosters was handed off to an "advanced program" using 437 technologies. This project, 437AP (for "alternate payload"), was a space-based reconnaissance system using photographic technology to perform a mission previously named SAINT ("SAtellite INspecTor").[86] The 437AP vehicle would use a Thor IRBM with a modified General Electric Mark II reentry vehicle. The Mark II was described as being very similar to the cameras used by the CORONA reconnaissance satellites.[87] In the declassified Program 437 history, Dr. Wayne Austerman stated the characteristics of the 437AP system: "[437's purpose] was to demonstrate the 'feasibility of obtaining technical intelligence photos of orbiting objects using a non-orbital interception system.' The 437AP payload was to be completely interchangeable with the Thor booster."[88]

As early as May 1964, AFSC's Space Systems Division created the contracts for modification of the Thor boosters and the acquisition of the as-yet-undesignated payload. By the time August rolled around, a System Program Office (SPO) had been established for 437AP, very soon after the original 437 SPO had closed its doors. Inside the Glossary of Defense Acquisition Acronyms and Terms, a "System Program Office" is defined as "the office of the [437AP] Program Manager and the single point of contact with industry, government agencies, and other activities participating in the system acquisition process."[89]

In common terms, creating an SPO for a weapon system is a big deal in terms of money, resources and personnel assigned. Creation of an SPO also implies that the acquired system was going to remain in the inventory long enough to justify the expense and resources assigned to it. Bureaucratic wrangling saw 437X, as it was then designated, to be the follow-on to Program 706, a system designed to photograph enemy satellites, an idea that had been around since 1958.[90] Since Program 437 was still operational, additional funds were provided to the SPO to allow modification for 437AP missions.[91]

The first launch of 437AP occurred from JI on December 7, 1965. The system flight parameters

Cousin to CORONA?

Author Ted Molczan wrote an informative article on Program 437AP in January 2016. His analysis goes beyond what is printed in Air Force Space Command's official history of the program and offers some graphical depictions of 437AP orbital trajectories and comparisons to the CORONA spy satellites. For a dark corner in military space history, Molczan covers the topic extremely well. His article can be found at www.satobs.org/Program437AP/Program437AP.html.

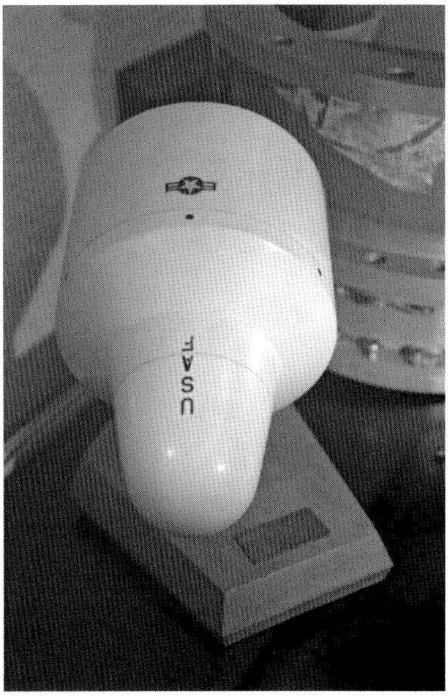

A model of the 437AP payload resides at the SLC-10 museum. Unlike CORONA taking pictures of areas on Earth, 437AP was designed to photograph other satellites. *Courtesy of the author.*

allowed photography of a target at an orbital altitude between one hundred and four hundred miles and a cross-range distance of eight hundred miles from its launch pad. The initial test used an Agena upper stage as a photographic target. According to aerospace historian Curtis Peebles, the photographic pass was a success, but the recovery capsule had a detrimental electrical problem preventing its separation and recovery.[92] The second 437AP launch took place on January 18, 1966, and was successful. Flight number three was launched on March 12, 1966. According to Dr. Austerman, the flight met "all research and development feasibility demonstration objectives."

Men and materiel used for 437 AP took away combat effectiveness for the 437 nuclear roles, as the dual booster synchronized launch processes could not be used if one Thor had a 437AP payload aboard. Time needed to swap out the AP payload for a nuclear warhead effectively eliminated the dual process. The last 437AP launch took place on July 2, 1966. The mission was to intercept and photograph NASA's Orbiting Astronomical Observatory (OAO-1), since it failed immediately after launch on April 8, 1966. However, an electrical short prevented the payload from imaging the target.[93] No further 437AP launches took place after this. Some fervor to launch another 437AP (to surveil a Soviet satellite) came from the commander of Continental Air Defense Command in March 1966 but was rejected since the location of JI had been revealed by President Lyndon B. Johnson to be an ASAT launch site. The 437AP program was cancelled by the Air Force on November 30, 1966.[94] Immediately after, personnel refocused back on the ASAT mission.

Phaseout of 437

On January 1, 1967, the 10th ADS was redesignated as a group, with two subordinate squadrons reporting to it: the 25th ADS at Vandenberg AFB and the 24th Support Squadron on JI. The 25th ADS was divided into three combat launch crews, which rotated to the island from the mainland on ninety-day tours. At Vandenberg, the 10th ADG took command of the DMSP satellite launch program previously conducted by the 4300th Support Squadron, which was inactivated on May 25, 1967. In his monograph "Shooting Down a Star," Dr. Clayton Chun describes one of the key events attributing to the downfall of Program 437: "[T]he shift of the nation's national security focus resulting from unfolding events in South Vietnam. Program 437's status as a DOD 'top priority' was lost forever after the Tonkin Gulf incident. Money, manpower, and other resources were quickly shifted to fight the Vietnam War."[95]

Ballooning costs from the Vietnam War saw many big-ticket space items cancelled as it became more likely the United States would be fighting terrestrial wars rather than space wars. The Manned Orbiting Laboratory (MOL), a spy satellite requiring a two-person crew, and Program 437 (and its variants) were not considered critical to the ground-war in Vietnam. MOL was cancelled in 1969, and Program 437 stayed on life support until the next half decade. Ironically, a "small-ticket" space item also launched by 10th ADS, DMSP, was critical to national security systems, aircrews and ground-pounders alike during Vietnam.

On December 31, 1970, the 10th ADG was downsized to squadron status, becoming the 10th ADS again. The two squadrons on JI, the 24th ADS and 25th ADS, were inactivated, and the personnel became part of Detachment 1, 10th ADS. The joint chiefs of staff had already ordered the system to stand down from twenty-four-hour alert in October, placing the system in a thirty-day recall status. Launch crews were sent back to Vandenberg to assist in other 10th ADS activities (e.g. DMSP), while a small contingent of caretakers remained on the island. The nuclear warheads were shipped to Nellis AFB, Nevada, for storage and recall, if necessary. The system continued to be neutered one subsystem at a time. The two on-alert missiles were sent back to Vandenberg. The ground guidance station was deactivated, and the LOX tank was drained. While the pieces remained intact, disintegrating the system caused reaction time estimates to increase.

Other happenings on JI around this time started changing the focus of the base's primary mission. In 1971, the Johnston Island Chemical Activity began stockpiling chemical weapons for eventual dismantling and destruction.[96] In mid-August 1972, Hurricane Celeste struck the atoll, causing millions of dollars in damage. Personnel from JI were evacuated to Hawaii a day prior; however, the hurricane damage proved to be problematic.

According to damage assessment by UNIVAC engineer Vernon Sandusky, "The storm did no damage that cannot be repaired. The 10 ADS has two problems however, not enough people to clean up the equipment, and lack of mechanical spare parts to replace units corroded beyond repair."[97]

After the hurricane damage was fixed, the unit returned to its footing of thirty-day recall. Aerospace Defense Command (ADC) cancelled the 1972 Operational Readiness Inspection, a requirement for combat-ready units. Administrative changes moved host-base responsibilities from the Air Force to the Defense Nuclear Agency in July 1973. Once the Air Staff ordered that the Program 437 nuclear warheads be retired on April 1, 1975, the system's demise was complete. JI hosted two more Thor launches in 1975, supporting ballistic missile defense testing, before the launch emplacements became silent.

The DOD stayed away from obtaining or retaining an antisatellite capability for the next decade, nuclear or conventional. The 1970s saw a change in focus from employing massive nuclear forces to developing strategies and treaties regarding nuclear armament control. The Strategic Arms Limitations Talks (SALT I) agreement (signed in 1972), the Anti-Ballistic Missile Treaty (also signed in 1972) and the SALT II talks (signed in 1979 but never ratified by the U.S. Senate) all required means of verification to ensure neither country was violating the treaty. In the United States, these capabilities were called National Technical Means (NTM) of Verification; the term NTM usually referred to spy satellites or airborne reconnaissance platforms. Speaking at Kennedy Space Center in Florida at the 1978 Congressional Space Medal of Honor Awards Ceremony, President Jimmy Carter said, "Photoreconnaissance satellites have become an important stabilizing factor in world affairs in the monitoring of arms control agreements. They make an immense contribution to the security of all nations. We shall continue to develop them."[98]

An ASAT capability, conventional or nuclear, would have threatened low-flying NTMs and undermined the good faith represented by the treaties. The United States chose to shelve its ASAT system. The Soviet Union, however, did not. It maintained its ASAT capability until the early 1980s.

Chapter 5

AMERICA'S EYES IN THE SKY

A generation of this Nation's youth has grown up unaware that, in large measure, their security was ensured by the dedicated work of your employees. National security interests prohibit me from rewarding you with the public recognition which you so richly deserve. However, rest assured that your accomplishments and contributions are well known and appreciated at the highest levels of our Nation's government.
—President Ronald Reagan, in a commendation to members of the GAMBIT program[99]

Author's note: While the National Reconnaissance Program (NRP) had many successes with its CORONA, ARGON, LANYARD, GAMBIT and HEXAGON series of photographic reconnaissance satellites, the launches took place at "sister" launch complexes of SLC-1, SLC-2, SLC-3 and SLC-4 at Vandenberg AFB. The dire importance of detecting cloud cover prior to launch of NRP photographic reconnaissance satellites is highlighted in chapter 5, focusing solely on DMSP. The critical nature of aerial and spaceborne reconnaissance during the Cold War, and its reliance on vital DMSP weather data necessitated a brief introduction in this chapter.

One of the Space Race's most defining moments didn't even occur during the era; it came nearly two decades earlier during the lead-up to World War II. The surprise attack on Pearl Harbor on December 7, 1941, would define the U.S. approach toward space and intelligence gathering for decades

The famed U-2 high-altitude reconnaissance aircraft, a critical part of the National Reconnaissance Program. *Courtesy of the National Reconnaissance Office.*

to come. The combination of a "Pearl Harbor–like" attack with nuclear weapons caused national leadership to place early warning and surveillance systems at the top of the defense priority lists. When the Japanese navy struck the fleet along Battleship Row on that sleepy Sunday morning in December, the initial effects of shock threw the United States into the largest movement of military might during the twentieth century. The secondary and tertiary effects—development of the atomic bomb and an unsettling feeling of vulnerability—defined the fears of a generation leaving a "hot war" and forced the nation into a "cold one."

For a nation that had just finished a war ending with the detonation of two atomic weapons, the immediate concern was keeping those same atomic secrets out of the hands of actual and potential adversaries. When the Soviet Union detonated its first atomic (fission) weapon on August 29, 1949, no one could have predicted the rapid growth of its nuclear weapons establishment. Only four years later, the country detonated its first hydrogen (fusion) bomb, code-named JOE-4.

The North Korean invasion of South Korea on June 25, 1950, also caught American intelligence unaware and (coming less than a decade after Pearl Harbor) escalated fears of a surprise attack.

DEVELOPMENT OF SKY SPIES

As early as 1946, supporters of aerial reconnaissance recognized the need for a strategic reconnaissance program to look at the "big picture," not just the tactical one presented by the adversary. A document prepared by the Army Air Forces, titled *The United States Strategic Bombing Survey*, chronicled the air war against Nazi Germany. One of the pioneers of national reconnaissance, Lieutenant Colonel Richard Leghorn, read a copy of the bombing survey and was appalled at some of the findings. Proper nodal analysis of traffic and electrical systems had not been performed, lessening the effectiveness of the massive attacks. Wartime concentration on tactical reconnaissance had given the Allies a soda-straw view of the battlefield. Leghorn, speaking at a dedication ceremony for the Boston University Optical Research Laboratory, later spoke of the need for strategic reconnaissance: "The nature of atomic warfare is such that once attacks are launched against us, it will be extremely difficult, if not impossible, to recover from them and counterattack successfully. Therefore, it obviously becomes essential that we have prior knowledge of the possibility of an attack."[100]

Running concurrent to the Thor IRBM program was another classified effort that was held at the highest levels of national security. In the early days of the Cold War, leaders were convinced an enemy would spring a surprise attack, a "nuclear Pearl Harbor," on the U.S. mainland. Warning systems, in addition to retaliatory means, were placed high on the defense establishment's acquition list. As millions of dollars were spent placing radar warning systems in the Arctic, members of the intelligence community began work on another type of warning system: a high-altitude photographic one—in essence, a craft that could fly over and "spy" on other countries. In the early 1950s, similar missions were being performed by aircraft such as the RB-36, RB-47E and RAF Canberra. These specially outfitted aircraft would fly "ferret" missions along hostile territory to activate and record electronic emanations from radars and surveillance systems. These recordings would be analyzed by the NSA and turned into signals intelligence (SIGINT) to assist organizations like SAC during their targeting operations. Collection of images presented a different problem. Since photographs taken from side views while on the periphery of hostile territory were not very good (higher altitude was needed to "see" farther inland), planners began developing ways to fly over specific areas to gain images.

One of the earliest efforts was the sensitive intelligence (SENSINT) program. President Eisenhower authorized this program in the fall of 1953 to allow the joint chiefs of staff and the director of Central Intelligence to request special overflight missions. One particular SENSINT mission, on April 14, 1956, took RB-47E photoreconnaissance aircraft from Thule Air Base, Greenland, to the Soviet city of Noril'sk.[101] The city was known only by a name and set of geographic coordinates, having been founded in the 1930s by Josef Stalin and declared off-limits to the rest of the world. The trio of aircraft flew at altitudes limiting their presence-revealing contrails, photographing Noril'sk, Igarka and the Port of Dudinka, before flying back over the North Pole to Thule.[102] Flight instruments informed aircrews that the Soviet radars had not detected them. This luck would not continue during later overflights. In 1956, Eisenhower ended the SENSINT program in favor of another reconnaissance aircraft, code-named AQUATONE—the venerable U-2 spyplane.

Another means of collecting overhead intelligence in denied territory was by use of high-altitude balloons. Recommended by a RAND Corporation study, the balloon idea used a lightweight plastic envelope to lift a camera gondola along high-altitude wind currents.[103] One balloon program, GENETRIX, ratched up 516 launches from December 1955 to February 1956. The slow drift gave a long transit time over the Soviet Union (around eight to ten days), allowing fighter aircraft plenty of time to shoot down the payloads.[104] When payloads succeeded in being recovered, the photos were a random mix of whatever landmass the camera flew over. The flights angered the Soviets, who made diplomatic protests during each wave of launches. When GENETRIX ended, only forty-six balloons were recovered.[105]

The U-2 spyplane was born from the idea that a plane that could fly high enough to avoid air defenses, such as intereceptor jets and surface-to-air missiles. The U-2 was nothing more than a powered glider with a thin fuselage and very long wingspan. The objectives behind the U-2 program were:[106]

- Providing adequate locations and analyses of Russian targets, including those newly discovered
- More accurate assessment of Soviet order of battle and of early warning indicators, thus improving our defenses against surprise attack
- Appraising Soviet-guided missile development (through photos of test range, etc.)

- Improving estimates of Soviet ability to deliver nuclear weapons and of their capacity to produce them
- Disclosing new developments that might otherwise lead to technological surprise
- Appraising Soviet industrial and economic progress

The U-2 flights over the Soviet landmass began on July 4, 1956. The incursions into their airspace did not go unnoticed by the Soviets. While the air defense fighters could not reach the altitude of the U-2, every flight was monitored. Diplomatic complaints, known as demarches, were forwarded from the Soviet delegation to the Americans. The risk of angering the Soviets was deemed acceptable since the photographic (and presumably signals) intelligence gained by each overflight peeled away the mysterious veneer of the Iron Curtain.

Just as the Americans were developing IRBMs and ICBMs, the Soviets tested their rockets from test ranges deep inside their territory. After the first test of the R-7 ICBM (NATO code name SS-6 SAPWOOD) on May 15, 1957, President Eisenhower allowed more overflights of Soviet territory. Photographs of the rocket test sites were analyzed by CIA experts to see how far Soviet rocket technology had progressed. The Americans, and the world, received the answer on October 4, 1957. Sputnik 1, the world's first artificial satellite, was launched from Tyuratam (now known as Baikonur Cosmodrome) into a low-Earth orbit.

While the R-7 rocket's unveiling to the world was as a space launch booster, CIA analysts were looking for evidence that it was being operationalized by the Russians. In 1959, the Strategic Rocket Forces (RVSN, *Raketnye voyska strategicheskogo naznacheniya*)

IMINT Versus PHOTINT

In the early days of aerial reconnaissance, cameras required film to record the images, hence the term photographic intelligence (PHOTINT). The manned platforms, like the U-2, would have to land after the mission to deliver the film for developing. Early first- and second-generation photographic reconnaissance satellites relied on similar film-based systems that used a parachute recovery to get the film back to Earth. With the launch of the first KH-11 electro-optical reconnaissance satellite in 1976, film was no longer required. Images were recorded digitally and beamed down to a ground station for delivery. The term imagery intelligence (IMINT) covers visiual photography, infrared sensors, electro-optics and synthetic aperature radar systems, where images of objects are reproduced on film or electronically.

was organized to control all intermediate-, medium- and intercontinental-range ballistic missiles. The CIA and military forces needed accurate intelligence to determine the capabilities and locations of future ICBM sites. In the political arena, American presidential candidates were arguing that the Soviets had technological superiority with their missile fleet and a missile "gap" existed in their favor.

> [CIA director] *Dulles was determined to obtain permission for more overflights in order to settle the missile-gap question once and for all and end the debate within the intelligence community. To accomplish this, Dulles proposed photographing the most likely areas for the deployment of Soviet missiles. At this time there was still no evidence of SS-6 ICBM deployment outside the Tyuratam missile test range. Because the SS-6 was extremely large and liquid fueled, analysts believed these missiles could only be deployed near railroads. Existing U-2 photography showed railroad tracks going right to the launching pad at the test site.*[107]

More overflights were authorized by President Eisenhower in the first months of 1960s. One flight in particular, Operation GRAND SLAM, took off from Peshawar, Pakistan, on May 1, 1960. The goal of GRAND SLAM was to fly across the Soviet Union from south to north, covering strategic targets at Tyuratam, Sverdlovsk, Kirov, Koclas, Severodvinsk and Murmansk. While the planned landing was to take place at Bodo, Norway, fate intervened. Four and a half hours into the planned mission, the U-2, piloted by Francis Gary Powers, was shot down by an SA-2 surface-to-air missile seventy thousand feet over Sverdlovsk. The ensuing political quagmire ignited by the shootdown saw the Eisenhower administration reveal its spying over the Soviet Union and place American intelligence collection in jeopardy.

World-Circling Spaceship

It took four years of U-2 flights (1956–60) for one to finally fall prey to the aircraft's vulnerabilities to high-altitude missiles. Even before Francis Gary Power's fateful flight, an alternative idea of collecting intelligence from space originated in a 1946 paper, "Preliminary Design of an Experimental World-Circling Spaceship." The military reconnaissance possibilities are identified on the first page:

The location of Vandenberg AFB was critical to allowing insertion of CORONA satellites into a polar orbit to maximize coverage of the Soviet Union. *Courtesy of the National Reconnaissance Office.*

> *In this report, we have undertaken a conservative and realistic engineering appraisal of the possibilities of building a spaceship which will circle the earth as a satellite. The work has been based on our present state of technological advancement....Such a vehicle will undoubtedly prove to be of great military value. However, the present study was centered around a vehicle to be used in obtaining much desired scientific information on cosmic rays, gravitation, geophysics, terrestrial magnetism, astronomy, meteorology, and properties of the upper atmosphere.*

Around the same time of the U-2 development, the USAF Scientific Advisory Board provided an analysis of then-current technology, realizing that small, lightweight thermonuclear (fusion) bombs were possible. Delivery of such weapons would require a long-range rocket (e.g. ICBM). Just as the Soviets discovered with the R-7's "dual duty" as a space launcher, the USAF acknowledged that the same thrust needed to lob a thermonuclear warhead over the North Pole could be used to place a satellite into orbit. In 1954, the USAF published a requirement for Weapon System 117L (WS-117L), a family of space-based systems for collecting photographic, electronic and infrared intelligence. The WS-117 project meandered along for a few years but unknowingly ran afoul of political machinations. The Soviet protests to U-2 overflights caused President Eisenhower to rethink the approach to the "publicly announced" goals of the WS-117L program, namely military surveillance. Knowing that it would be hard to hide the intent behind a military surveillance satellite, the mission would have to be kept covert or unknown to the general public under the guise of a cover story. Failure of the GENETRIX balloons and U-2 flights to remain covert caused the impudent diplomatic Soviet backlash. Eisenhower wanted to avoid any further embarrassment regarding the U.S. intelligence needs.

In mid-1957, U.S. intelligence analysts believed the Soviets would launch a satellite before the year's end. As part of the plan to legally justify the overflight of a reconnaissance satellite, Eisenhower did not press the leaders of the Vanguard satellite program (vying to be the first American satellite in orbit) to beat the Soviets into space. The launch of Sputnik and its subsequent orbital path over the United States set a legal precedent that every satellite since has followed—unimpeded travel over the Earth, in stark contrast to any claims of sovereign "airspace" over a particular country.

Unique circumstances existed for monitoring a landmass at very high northern latitudes. The solution was a "polar orbit," allowing a satellite to circle over the North and South Poles. During a polar orbit, the Earth below rotates at a constant rate below the spacecraft, allowing viewing of every inch of the globe over time. Vandenberg's claim to fame was as the only site in the United States that could allow a launch into polar orbit without the boost rocket flying over a populated area. A launch from the California coast doesn't fly over another landmass until Antarctica.

While the technology for the WS-117L family of satellites progressed at a nominal pace, the most mature piece was "removed" from the public program and placed into covert channels.[108] The photographic return system, utilizing a camera, film reel and reentry capsule (also referred to as a "film bucket"), was reoriented into an experimental research program called Discoverer. This publicized change in focus allowed the Air Force to continue purchasing Thor boosters, Agena upper stages and reentry capsules for the purposes of "biomedical experiments." The only pieces that still had to be procured in secret were the photographic cameras and film, necessitating continued security on the classified aspects. The effort was given the code name CORONA and was run by a joint Air Force–CIA leadership team in a similar manner to the U-2 effort.

The first launch attempt of a Discoverer mission took place on January 21, 1959. One hour before launch, a routine test of the Agena's hydraulic system was initiated. Thirty seconds after initiation, the explosive bolts and separation rockets holding the Agena upper stage ignited. Luckily, the upper stage did not tip over and fall; however, corrosive oxidizer spilled down the side of the Thor and damaged the guidance system. Since lift-off did not occur, the mission was designated Discoverer 0 (zero).[109]

The trio of flights that followed were all R&D missions to bolster the Discoverer cover story. These flights unintentionally provided the best "you gotta be kidding me" anecdotes of the program. Discoverer 1 lifted off on February 28, 1959, from Vandenberg in full view of the public and press. Minutes into the flight, the Air Force tracking network lost contact with the satellite. Sporadic signals were retrieved, dashing any hopes that the satellite had arrived in the proper orbit. It was determined weeks later that the satellite did not obtain orbital velocity and reentered over the Antarctic. Unbeknownst to CORONA personnel at the time, it would take thirteen launches before the first wholly successful mission.

The saga of Discoverer 2 helped inspire a bestselling novel and subsequent Hollywood movie. After being launched on April 13, 1960, the Air Force tracked the satellite in a stable low-Earth orbit. The Agena upper stage successfully became the first three-axis stabilized satellite in orbit. When it was time to recover the reentry capsule, the release command was transmitted. Due to the lower-than-expected orbital altitude, the reentry capsule came in earlier than planned, over the country of Norway instead of the balmy Pacific Ocean. It was estimated to have landed in the area of Spitsbergen (now called Svalbard). Individuals from a nearby mining encampment reported seeing the reentry and a deployed parachute.[110] Lieutenant Colonel Charles "Moose" Mathison, commander of the 6594th Test Wing, contacted the Royal Norwegian Air Force to search for the wayward capsule. The search began on April 16 but was abandoned six days later.[111] Reports revealed that ski tracks were found near the impact area and led back toward a nearby Soviet mining camp. National Reconnaissance Office (NRO) historical documents stated, "The Discoverer II capsule might have been recovered by the Soviets after re-entering and returning to Earth on Spitzbergen Island, and fact that the Norwegian authorities may have provided credible evidence of that possibility."[112]

Discoverer 3 provided more unintentional humor before even lifting off the ground. The third mission was an attempt to maintain a plausible story for the Discoverer's biomedical experiments.[113] As originally conceived, the satellites would launch live specimens—mice and primates.[114] Additional payloads would include radiometric packages to measure the space environment and assist in the "other" satellite programs. Aboard Discoverer 3 was a group of four "astro-mice." Inside the reentry capsule, the mice sat in cages a little more than double their size with biomedical radio transmitters and a three-day food supply. During the first launch attempt on May 21, 1959, ground controllers were alerted to the mice's inactivity; the critters had apparently died after eating some spray paint applied to the inside of their capsule.[115] Nearly two weeks later, another launch was attempted. The second crew of mice had inadvertendly urinated on a humidity sensor, causing it to read "100% relative humidity." After the sensor dried out, the countdown was resumed. Discoverer 3 lifted off from Vandenberg AFB on June 3, 1959. Insufficient velocity prevented the satellite from reaching orbit.[116] The rocket, payload and "crew" reentered the atmosphere to a watery grave. Animal cruelty complaints were received by the U.S. governement, so the follow-on flights with live mice and a primate were abandoned.[117]

After the trio of R&D missions, the first CORONA camera would be launched on the next flight. Since the cover story was still intact, the mission was publicly designated Discoverer 4, while CIA-NRO records list the flight as CORONA Mission 9001. The 9000 series would designate the first generations of CORONA capsules; subsequent modification or distinct missions would garner a unique series number (e.g. KH-5 ARGON was 9000A, KH-6 LANYARD was 8000). This mission would not obtain orbit; the Agena upper stage would again fail to provide enough thrust to obtain escape velocity. Regarding the distressing losses, CIA program manager Dr. Richard Bissell described the feeling:

It was a most heartbreaking business. If an airplane goes on a test fight and something malfunctions, and it

When History Meets Hollywood

The launch and subsequent loss of Discoverer 2 in Norway inspired Alistair MacLean's Cold War yarn ***Ice Station Zebra***. In its Review and Redaction Guide, the NRO revealed that "an individual formerly possessing CORONA access was the technical adviser to the movie" and "the resemblance of the loss of the DISCOVERER II capsule, and its probable recovery by the Soviets" influenced the book ***Ice Station Zebra*** and subsequent movie. A remake of the film is currently in development.

> *gets back, the pilot can tell you about the malfunction, or you can look it over and find out. But in the case of a* [reconnaissance] *satellite, you fire the damn thing off and you've got some telemetry and you never get it back. There is no pilot, of course, and you've got no hardware, you never see it again. So you have to infer from telemetry what went wrong. Then you make a fix, and if it fails again you know you've inferred wrong. In the case of CORONA it went on and on.*[118]

While the CIA worked through the series of Discoverer/CORONA flights, the engineers started working down a list of solving problems on this radically new technological leap. Some solutions themselves caused problems that required troubleshooting. Overheating inside the satellite shroud was fixed on Discoverer 9 by a water-cooling apparatus, but sloshing water in the Thor caused movement on the rocket's center of gravity. The fix was an ingenious solution of sanitary napkin filling mixed into the water and worthy of mention in the official CORONA history:

> *A standdown was in effect from November 20, 1959 until February 4, 1960 to allow time for intensive R&D efforts to identify and eliminate the causes of failure. During this period of problem solving, one amusing and innovative design bears mention. A cooler was needed for the fairing interface which was heating up during ascent....In order to contain the water* [coolant] *and prevent sloshing, something absorbent, soft, and easy to work with was required. After conducting a test program on various materials, the design engineer chose "Modess because."*[119]

Failures with the missions continued, but strides toward total success were being made. On Discoverer 11/CORONA mission 9008, the Agena boosted to the correct orbit, the camera operated satisfactorially and the reentry vehicle detached from the satellite on command. Due to an unknown malfunction in the detachment spin rockets, the reentry vehicle did not enter the atmosphere. The next mission, Discoverer 12, was a diagnostic R&D mission without a camera to test the separation rockets. The flight was launched on June 29, 1960, but the Agena failed to attain the proper orbit.[120]

LUCKY THIRTEEN

Never Quite What It Seems
While the Discoverer experimental payload cover story held through the launch and recovery of "Lucky Thirteen," its payload was not just an American flag. Declassified in 2015 by the NRO, the AFTRACK program placed a SIGINT sensor near the rear of the Agena.[121] This sensor would detect if the Soviet Union's radars had tracked the satellite's orbit. By the launch of the thirty-eighth Discoverer on February 27, 1962, the biomedical experiments cover story had worn thin and the name was dropped.[122] All CORONA-related launches would remain unidentified until the program's declassification in 1995.

The next mission, Discoverer 13, would replace the diagnostic flight plan for the lost Discoverer 12. The vehicle, sans camera and film, lifted off from Vandenberg AFB on August 10, 1960. The Agena was properly inserted into orbit, and on the seventeenth revolution around the Earth, the reentry vehicle separated on command from the satellite. The capsule reentered over the Pacific Ocean, albeit farther away from the planned recovery area. As a backup to a failed aerial recovery, the capsules were outfitted with a radio transmitter to guide recovery forces during a water landing. If, in the case of a photographic mission, friendly forces were unable to recover the floating capsule, a salt-plug would dissolve and allow the film and capsule to sink to the ocean floor. This was not the case for Discoverer 13; the capsule was plucked from the ocean by U.S. Navy divers. Discoverer 13's payload, an American flag, was taken back to the United States for presentation to President Eisenhower on August 15, 1960.[123] The recovery was celebrated by the public as the first object returned from space and in the classified realm of the CIA for a successful recovery. The next launch, Discoverer 14, would be just as successful and would carry an important payload: the first intelligence photographs taken from space.

While CORONA continued on, the launches and recoveries became routine. All the trials and tribulations learned through the failures of the first launches paid off. Accidents still occurred, and some aerial recoveries dropped into the ocean, but overall, CORONA soldiered on to become one of the most successful satellite programs in space history. During the program's lifetime (1959 through 1972), CORONA would tally 144 missions, collecting over 800,000 images and covering over 510 million square nautical miles of the Earth.[124]

President Dwight Eisenhower inspects the first object returned from space, an American flag, as Air Force chief of staff General Thomas D. White and Colonel Charles "Moose" Mathison look on. *Courtesy of the National Reconnaissance Office.*

The list of "firsts" from the CORONA program is impressive:

- first photo reconnaissance satellite in the world
- first man-made object returned to Earth
- first midair recovery of a vehicle returning from space
- first mapping of Earth from space
- first stereo-optical data from space
- first multiple reentry vehicles from space
- first reconnaissance program to fly one hundred missions
- first reconnaissance satellite program to be declassified

ORIGINS OF THE NRO

The shared use of reconnaissance assets by the CIA and the Air Force in both the U-2 and CORONA programs produced some of the best intelligence seen. The effort also caused management headaches with infighting about mission resources and future programs. As Dr. Jeffrey Richelson states in his monograph, "Civilians, Spies, and Blue Suits: The Bureaucratic War for Control of Overhead Reconnaissance, 1961–1965," the NRO

> *was different from most other government organizations in two ways—its very existence was classified and its key components were actually elements of the Air Force, CIA, and Navy. In the five years following their creation, the NRP and NRO were the subject of intense battles between the CIA and the civilian and uniformed Air Force officers who ran the NRO. At first the battles primarily focused on the authorities of the NRO and its director. Subsequently, a major aspect of the conflict involved decisions concerning new reconnaissance systems.*[125]

Created in 1961, the NRO took on the responsibility for the National Reconnaissance Program (NRP), consisting of airborne and space-borne systems to monitor changes in the strategic balance between the United States and Soviet Union. Since the successful cooperation between the CIA and Air Force had existed since the U-2 program and continued through CORONA, it was decided that the new office would be a mixture of the two, with a sizeable supporting contractor force. Lessons learned by both the CIA and Air Force during CORONA, such as covert acquisitions of space hardware, were successfully applied to follow-on programs such as GAMBIT and HEXAGON:

> *The NRP will consist of all satellite and overflight reconnaissance projects whether overt or covert. NRP will include all photographic projects for intelligence, geodesy and mapping purposes, and electronic signal collection projects for electronic signal intelligence and communications intelligence resulting therefrom.*
>
> *There will be established on a covert basis a National Reconnaissance Office to manage this program. This office will be under the direction of the Under Secretary of the Air Force and the Deputy Director (Plans) of the Central Intelligence Agency acting jointly. It will include a small special staff whose personnel will be drawn from the Department of Defense and the Central Intelligence Agency. This office will have direct control over all elements of the total program.*[126]

The NRO also pushed for the development of a weather satellite to support its photographic reconnaissance operations. (This effort will be covered in the next chapter.)

NEXT GENERATION: GAMBIT

While CORONA was working out its "teething" problems in the early years, another film-based surveillance satellite was being developed. Eastman Kodak provided the government with documents on the feasibility of a space-based camera that could provide two- to three-foot resolution. In comparison, the first CORONA photographs could provide forty-foot resolution, which would whittle down to six-foot resolution by the end of the program. As told in the official history, *The GAMBIT Story*:

> *The definitive steps which led to the GAMBIT program were taken in early 1960—before the CORONA photographic reconnaissance satellite had its initial success. On 24 March 1960, the Eastman Kodak Company submitted an unsolicited proposal to the Air Force's Reconnaissance laboratory at Wright Field. This proposal…suggested development of a high-performance, 77-inch focal-length, catadioptric-lens camera suitable for satellite reconnaissance.*[127]

After the successes of the U-2 program, the needs of the intelligence community dictated that technical intelligence be considered the next desirable goal. Instead of counting concrete launch pads with a twenty-five-foot resolution photo, higher-resolution photographs could show numbers of support equipment trucks on the launch pad. A short introduction from *The GAMBIT Story* describes it best:

> *Traditionally, experts in analyzing reconnaissance photography functionally divide it into two categories. One is called "search," and is dedicated to answering the question, "Is there something there?" CORONA's camera was designed to photograph large contiguous areas in a single frame of film in order to provide answers to that question. Even though CORONA's resolution improved from its original 35–50 feet to 6–10 feet, its basic function remained search.*[128]

The second category identified is called "surveillance" and is used after something interesting has been found. After discovery of a target of interest, intelligence analysts learn more in the attempt to identify and classify it. The GAMBIT series was designed to be a high-resolution "spotter" for areas of interest that CORONA had already identified.

In a similar vein to the WS-117L project "cancellation" and then transition into covert channels, GAMBIT's development ran parallel to the SAMOS E-6 satellite program. To even further distance the origins from the WS-117L's military reconnaissance lineage, GAMBIT program managers played complex shell games by creating a "null program," one with no stated goals and no origin to draw conclusions from. This allowed GAMBIT program managers to purchase equipment without snooping parties tracing the effort back and no cover story to run thin as time went on.

The GAMBIT family consisted of two satellites: the KH-7 GAMBIT-1 and the KH-8 GAMBIT-3. The KH-7 consisted of a fifteen-foot-long, five-foot-diameter satellite carrying a camera that could provide two- to three-foot resolution from a nominal ninety-mile altitude. Photos were taken on a 9.5-inch-wide film over a three-thousand-foot spool. With the increase of camera weight, film and reentry vehicle size, a larger booster was used. Just like the Thor before it, the Atlas ICBM was repurposed into a space booster, complete with an Agena-D upper stage. A unique change from the CORONA design, aside from the larger camera, was the creation of an orbital control vehicle (OCV) that allowed more accurate pointing throughout a mission profile.[129] As the satellite flew in its orbit, the OCV would slew the camera back and forth, aiming at targets beneath and along the satellite's path.

> *Pointing had to be extremely precise, requiring extreme accuracy in the horizon sensors and a stable platform gyro system that would allow the sensors to stay locked on the horizon while the vehicle rolled to point toward targets on either side of the orbital track. Because the ground swath width of Gambit cameras was only 10 miles, more photographs would be taken from a canted than from a vertical position.*[130]

Unlike the problems that beset CORONA's early flights, the first GAMBIT-1 flight from Vandenberg AFB on July 12, 1963, was successful. The direction given to Colonel William King, the GAMBIT program director, was to get "one good picture" from this mission. Ninety minutes after launch, the GAMBIT mission 4001 was found in a proper polar orbit

110 nautical miles above the ground. After nine orbits, the reentry vehicle was ejected back to Earth. When it was retrieved, the capsule held 198 feet of film—a minor use of the nearly 3,000 feet of film provided. The photos validated the camera, with photos ranging from 10-foot resolution down to an inconceivable 3.5-foot resolution.[131]

As the GAMBIT-1 flights progressed, an advanced version was proposed. The GAMBIT-3 (also known as the KH-8) had requirements of:

- High reliability
- First flight by July 1, 1966
- Ground resolution of [redacted] contrast, as presented to the front aperture of the camera in black-and-white film for vertical daylight photography at a payload altitude of [redacted]
- Day or night photographic coverage of targets. Exposure commandable on orbit from a selection of exposure steps
- Use of black-and-white film, color film and/or other special recording materials
- Stereo, strip and lateral pair photography
- Dual-camera subsystem configuration
- Minimum mission life of eight days
- Utilization of Atlas/Agena-D launching vehicle, or equivalent

Stringent requirements for better resolution (on the order of inches) required a more evolved camera and more weight added to the system. The Atlas-Agena combination was dropped in favor of the Titan-Agena combination. The first twenty-two flights of the GAMBIT-3 missions flew aboard Titan IIIB/Agena combinations. Later GAMBIT-3 versions flew aboard Titan 23Bs and Titan 24Bs, variants of the original Titan IIIB.

The spotter/shooter combination of CORONA and GAMBIT gave an awesome intelligence collection capability to the United States. CORONA's search area capability allowed targeteers to provide locations for GAMBIT operators to hone in on:

> *The CORONA program provided, for the first time in US history, a capability to monitor military and industrial developments over vast areas of the Soviet Union and other denied areas of the world. Although CORONA provided immeasurable contributions to national security, its resolution was not good enough to answer numerous critical intelligence questions, such as those regarding weapons development that the United*

States needed to guide counter weapons development. Nor could it provide the image quality the scientific and technical (S&T) intelligence organizations required to do true S&T analysis. GAMBIT aptly filled this high-resolution need and, by the end of the program, was routinely collecting [high-resolution] *imagery.*[132]

End of GAMBIT

Even with amazing resolution provided by its mirrors, occasional GAMBIT missions would provide mysteries that would befuddle even the most experienced photo interpreters. One particular flight in 1984, Mission 4354, flew over a Soviet special research and development facility in the Caspian Sea and photographed a large vessel that had not been seen previously.[133] Part boat and part aircraft, the enigma was designated the "Caspian Sea Monster." GAMBIT took such exquisitely detailed photographs that analysts were able to create engineering drawings of the craft. What remained hidden was the purpose for the monster's creation. After the end of the Cold War, the monster's design and purpose was revealed: it was a surface-effect vehicle—in effect, a seaplane that flew above the water to transport troops and materiel.

When the CORONA program ended in 1972, GAMBIT's surveillance capabilities were paired with the KH-9 HEXAGON's search. The GAMBIT family ended its program run in 1984. GAMBIT-1 (KH-7) had a total of twenty-eight successful missions out of thirty-eight attempted.[134] GAMBIT-3 (KH-8) ended its run with fifty successes out of fifty-four missions launched.

Cadillac in the Sky: HEXAGON

When CORONA hit its stride in the first half of the 1960s, the CIA commissioned a study in 1964 on a new search system, known as FULCRUM.[135] While CORONA was meant to serve as an interim search system, its success allowed the program to continue until follow-on systems could be designed and built. One idea to combine goals and funding was to produce a satellite that could do search and surveillance. This new system, called HEXAGON (also known as KH-9 or colloquially as "Big Bird"), would provide "an optimum capability for fulfilling the search and surveillance objectives specified...[s]ystematic search of some 12 million square nm semiannually...[coverage] during periods of crisis [with] minimal time between launching, recovery, and delivery of photography to the user."[136]

What U.S. intelligence analysts wanted was akin to a space truck—carrying wide-area search and close-look surveillance cameras, able to photograph areas of interest during routine and crisis situations and still having reserve capability on orbit. During the late 1960s, two rapidly developing crises began *and* ended during a time period between CORONA and GAMBIT launches: the Six-Day War in 1967 and the Soviets' invasion of Czecholovakia in 1968. Indications and warning clues were not recognized in a timely manner to allow launches of surveillance satellites to monitor the situation. HEXAGON's design included four film reentry capsules. The extra film allowed the satellite to routinely monitor identified areas, and in the case of a rapidly developing situation, the photographs would be returned to earth as soon as possible. The satellite's lifetime would last as long as the film held out.

From the Depths, to the Depths

During HEXAGON's inaugural flight, the third reentry vehicle was not snatched during aerial recovery, nor was it grabbed during the water-borne landing. The over one-thousand-pound object sunk to the bottom of the Pacific Ocean, to a depth of nearly 16,400 feet. A sizable recovery operation took place weeks after, but the film disintegrated. The tale of the recovery effort is captured in the CIA monograph "An Underwater Ice Station Zebra: Recovering a Secret Spy Satellite Capsule from 16,400 Feet below the Pacific Ocean."

This high-resolution image of the U.S. Capitol building in Washington, D.C., was taken by KH-7 Mission 4025 on February 19, 1966. *Courtesy of the National Reconnaissance Office.*

Placed into orbit on June 15, 1971, HEXAGON's first flight (Mission 1201) provided the best of both worlds:

> *While the primary objective of the HEXAGON mission was to provide high resolution photography over broad areas, the intent of the first flight was to demonstrate functional operation of the system. The sensor system certainly achieved this intent. One of the* [photographic interpreters] *at the Eastman Kodak processing facility remarked, "My God, we never dreamed there would be this much, this good!"*[137]

The HEXAGON program would provide film-based imagery through a total of nineteen successful flights out of twenty launches. The last launch, Mission 1220, provided a bittersweet end to the program: "The HEXAGON flight program ended sadly. On April 18, 1986, the launching of the last vehicle was terminated by a catastrophic booster failure nine seconds after liftoff."[138]

DEAD ON ARRIVAL: DORIAN

Mentioned here as a footnote in national reconnaissance history, the KH-10 DORIAN (also known as the Manned Orbiting Laboratory or MOL) was the melding of a photographic reconnaissance satellite with a manned spacecraft. This system was envisioned to be more responsive to policymakers' needs by allowing a "human-in-the-loop" to make decisions on when and where to take photographs. Astronauts would fly a modified Gemini capsule to the MOL and stay aboard on a set periodic schedule. When operating the KH-10 camera/telescope, the astronaut could slew the camera toward areas of interest, as directed by ground controllers, and decide whether to take the photograph or not (if cloud cover prevented capture). MOL ended up on the budgetary chopping block in 1969, at the height of spending during the Vietnam War. The success of the civilian Apollo moon program also put the nail into the coffin, as justification for two astronaut programs could not be discussed outside classified environments.

Declassified in 2015, documents on the MOL's design showed two modes of satellite operation: manned and unmanned. In the manned configuration, the two astronauts would photograph targets and return all film at the end of the mission. The unmanned configuration allowed for a sixty-day mission with six independently released reentry vehicles.[139] In comparison, HEXAGON carried less film but had longer on-orbit times.

ONE STANDS ALONE: QUILL

Hidden within the 145 missions under the CORONA program banner was an outlier—one satellite that looked like the rest but was radically different. The program's name was QUILL, and it was the world's first imaging radar satellite. Photographic reconnaissance satellites record their targets onto film; the critical requirement is light—sunlight to illuminate the target area. Things that obscure the target include clouds and darkness. The use of radar waves to "paint the picture" allowed the operators to create an image of the area of interest using reflected energy from the satellite or aircraft. The technology had been scrutinized in the 1950s as a possible payload for the ill-fated WS-117L program, along with photographic cameras and infrared sensors. The utility of early radar images was posted on the front page of the *Washington Post* on April 20, 1960, when the U.S. Army imaged American cities at night and through cloudy weather from a small aircraft.

In his monograph about the QUILL satellite, Robert L. Butterworth states, "After talking with officers of the Strategic Air Command (SAC), [QUILL engineers] concluded that radar with about 10-foot resolution would be useful for post-strike bomb damage assessment, particularly because it could respond quickly and not have to wait for clear skies and sunshine."

The purpose of SLC-10's DMSP launches was primarily to provide meteorologic coverage for the military and NRP. The rendering of QUILL's products seemed, at first glance, to make the weather satellite moot. The experiment, however, was more successful in proving how important DMSP's information was to all reconnaissance satellites, not just the IMINT programs. Atmospheric moisture and rainfall were culprits from which neither photographic nor radar images could escape. A passage from the QUILL evaluation report showed what returned to the satellite's sensors through a rainstorm:

> *Intense rainfall was occurring locally in the area.... While the clouds themselves are generally not imaged, the occasional regions of dense rain scatter considerable signal back to the radar, causing cloud-like forms in the image. Inspection of the film shows that the ocean wave-structure patterns can be observed even in the densest parts of the rain-returns. Furthermore, although the raindrops both back-scatter and attenuate the radar waves, hence reducing the illumination at some greater slant range, no shadows of these rain cells have been noticed in the imagery.*[140]

The QUILL program officially ended on February 10, 1969, when the name was discontinued under the BYEMAN Control System.[141] The results gathered from the experiments were renamed "Radar Feasibility Study" and were not declassified and released to the public until 2012.

NEXT-GEN TECHNOLOGY: KH-11

As mentioned earlier, the two rapid crises of the 1967 Six-Day War and 1968 invasion of Czechoslovakia proved that the film-based photoreconnaissance satellites had a serious limitation: timeliness. The crises had already ended by the time a satellite could be launched. Development of a TV-relay imaging satellite was undertaken during the early SAMOS effort under WS-117L but was abandoned. Not until the invention of the charged-coupled device (CCD) in 1969 was the concept of a near–real time reconnaissance satellite realistic. No longer would satellites require film supplies to return to earth for development and analysis. Imagery, in digital form, would be transmitted to a ground station in minutes and hours instead of days and weeks.

Through a storied development program that still remains classified, the first near–real time electro-optical intelligence satellite, named the KH-11, was launched on December 19, 1976, from Vandenberg AFB. The system was declared operational by President Gerald Ford on January 20, 1977.[142] Even though GAMBIT and HEXAGON were still in operation for another decade, the writing was on the wall for film-based reconnaissance satellites.

CLEAR "EYES" MAKE ALL THE DIFFERENCE

William Burrows's 1986 exposé, *Deep Black: Space Espionage and National Security*, starts with a bang. During a routine political stop during his reelection campaign, President Lyndon B. Johnson met with government officials and teachers on March 15, 1967. L.B.J.'s comments on the "space program" were quite telling:

> *We've spent thirty-five or forty billion dollars on the space program. And if nothing else had come of it except the knowledge we've gained from space photography, it would be worth ten times what the whole program cost. Because tonight we know how many missiles the enemy has and, it turned out, our guesses were way off. We were doing things we didn't need to do. We were building things we didn't need to build. We were harboring fears we didn't need to harbor.*[143]

At this point in the space reconnaissance timeline, the CORONA program had flown 115 flights. The close-look satellite family GAMBIT was in full swing. The KH-7 GAMBIT-1 satellites had flown 36 flights, with 2 remaining before the program's completion in the summer of 1967. The follow-on, KH-8/GAMBIT-3, had only 4 flights completed by LBJ's speech but continued on its predecessor's successful legacy.

While the focus of this chapter was primarily on reconnaissance satellites flown from Vandenberg AFB's other space launch complexes, the critical limiting factor on *all* of these systems—from CORONA to GAMBIT and HEXAGON to the KH-11—was weather. Thick cloud cover would kill any mission attempting to photograph the Soviet Union. Even with advanced technology systems, this lesson remains true today.

TABLE 3. NATIONAL RECONNAISSANCE PROGRAM ("KEYHOLE") SATELLITES, 1959–1986

Keyhole Designation	Camera	Number Launched	Time Period
KH-1	C	10	1959–60
KH-2	C' (C Prime)	10	1960–61
KH-3	C''' (C Triple Prime)	6	1961–62
KH-4	M (Mural)	26	1962–63
KH-4A	J (J-1)	52	1963–69

Keyhole Designation	Camera	Number Launched	Time Period
KH-4B	J-3	17	1967–72
KH-5	--	12	1961–64
KH-6	E-5	3	1963
KH-7	--	36	1963–67
KH-8	--	54	1966–84
KH-9	--	19	1971–86
KH-10	--	0	Developmental; never flew
KH-11	--	(classified)	1976–??

Chapter 6

THE FIX TO A CLOUDY PROBLEM

CORONA Obscura

Prior to the launch of the first CORONA missions, the Rand Corporation had given warning to the U.S. Air Force that systemic photographic coverage of the Sino-Soviet landmass would be greatly hindered by clouds. The autonomous operation of CORONA's cameras did not allow for operator input to turn the system on and off when the weather was bad. Another concern was that a large portion of the small film load on early CORONA satellites would be obscured. Any shots of cloudy landmass were wasted, and the photo interpreters on the ground would not know the film's disposition until the reentry capsule was recovered. Their fears were well founded. In fact, almost half of the film collected in the early days of the program (1959–61) had images of clouds. A reliable and timely way to detect the meteorological conditions of the Soviet Union prior to a CORONA launch was desperately needed. In a document written for WS-117L titled "Weather Factors in Satellite Reconnaissance," the need for a dedicated meteorological satellite is outlined, as are some obstacles:

> *The effect of weather on photo reconnaissance has been known since aerial reconnaissance was first tried, but a set of new elements is brought about by the use of satellites.*
>
> *1) Since satellites are outside the atmosphere, clouds at all altitudes can prevent photo reconnaissance thereby increasing "effective" cloud cover.*

> *2) Surveillance from one polar satellite at about 300 miles gives successive looks at the same location at roughly 24-hr interval. The relation between this interval and the "correlation interval" for cloud cover is not too well known, but plays an essential part in operational plans for photo surveillance.*
>
> *3) The usual weather computations for satellite photo surveillance define a necessary satellite-day figure. The implication that the same probability of obtaining photo coverage is obtainable by using one satellite and trying for 15 days, or using 15 satellites and trying for one day, is obviously false. Here the satellite-day figure is optimistic because of the observation made under* [the assumptions] *above.*
>
> *4) Generally speaking, at higher latitudes the weather gets worse but the overlap between orbits is high; the two effects tend to compensate* [for] *each other and to equalize at all latitudes the number of satellite-days required to give a certain probability of success for photo reconnaissance.*
>
> *5) In some parts of the world, the weather is so bad that statistical success in photo surveillance becomes so unlikely as to destroy the usefulness of the photo mission. In this case, all-weather sensors and directed reconnaissance, rather than surveillance, becomes the only possible solutions.*[144]

A 1961 interdepartmental study gave NASA the responsibility of developing a shared meteorological satellite for the Department of Defense and Department of Commerce, designated the National Meterological Satellite Program. Initially, the NRO had hoped to use data from NASA's TIROS (television infrared observation satellite) weather satellites, expected to be in orbit by 1960, and the later NIMBUS program. However, two main concerns with using the civilian weather system arose: political implications from using a "peaceful" space system for reconnaissance and timeliness of the information run by a civilian agency. Another concern was operational security (OPSEC) of the NRP's existence:

> *If reconnaissance authorities did intrude* [on the civilian weather satellite programs], *there was the danger of public protests that would in effect advise the Soviets that the United States needed weather information (chiefly cloud cover data), and hence presumably was operating reconnaissance satellites. In the early 1960's, those operations had gone underground and there was no immediate prospect of their surfacing in the near future.*[145]

A technician inspects the launch shroud before encapsulating the DMSP satellite (lower left). Note the white scrubs and clean room accessories for satellite processing. *Courtesy of Vandenberg Space and Missile Technology Center.*

On June 21, 1961, the decision was made by DNRO Joseph Charyk to develop an "interim" meteorological satellite program, with the first launch in ten months. Lieutenant Colonel Thomas O. Haig, the first director of the Defense Meteorological Support Program, took on the challenge delegated by Charyk. Knowing that resident systems engineering and technical direction contractors thrive on changes to satellite systems for their financial livelihoods, Haig took the unconventional step of *not* using one. Additionally, Haig fought for fixed price, fixed delivery contacts, choosing his own staff and complete and direct control over the program—very radical deviation from the standard aerospace contracting process but changes that would ultimately prove successful. His superiors agreed, due to the critical need for such a system, and added an additional contract "exit" point—if the launch schedule for the first satellite was not met or there were cost overruns over the provided fixed budget, Haig's superiors instructed him to terminate the contract and recover what funds were left.[146]

An example of the hard-nose tactics used during the program is highlighted in R. Cargill Hall's monograph on DMSP:

> [The] *fixed-price, fixed-delivery contract proved itself in December 1961, when a major structural member of the weather satellite, the base plate, failed during tests and company officials requested a three-month delay for redesign....After discussion* [the government] *advised RCA that it had ten days to produce a fix or the contract would be terminated under procurement regulations "at no cost to the government." The RCA program manager appeared three days later with revised internal schedule that met the original launch date.*

The compressed timeline for system acquisition and launch gave very little leeway. The satellite did not contain any redundant equipment, a necessary feature for a system in a hostile space environment. The "high-risk" status of the system illustrated a design that was "throwaway," a single purpose at minimum cost. This perception of a program with an unknown chance of success gave opportunities to incorporate innovative design features. One such feature was the spin-axis held perpendicular to the Earth's magnetic field via a direct-current wire loop around the satellite bus. This technique, championed by Lieutenant Colonel Haig and Lieutenant Ralph Hoffman, kept the satellite sensors aimed toward the Earth without relying on additional thrusters or on-board fuel. Another innovative sensor provided by civilian meteorologist Dr. Verner Suomi collected temperature variances from two one-inch squares of aluminum. Satellite analysts soon realized this data allowed the prediction of cloud cover during nighttime passes. Suomi's sensor design became standard on DMSP satellites until the advent of the operational

10th Aerospace Defense Squadron technicians inspect the DMSP payload shroud on a Thor Burner IIA rocket. *Courtesy of Vandenberg Space and Missile Technology Center.*

This Thor-Burner II launched from the SLC-10W pad on October 11, 1967, lifting a DSMP Block 4A satellite into polar orbit. *Courtesy of Vandenberg Space and Missile Technology Center.*

line scanner (OLS), which provided images of cloud cover in both visual and infrared spectrums.

The satellite's design was derived from TIROS but smaller: a ten-sided polyhedron that was twenty-one inches high and twenty-three inches across. The primary on-board sensor was a vidicon that could record an 800-square-mile picture of the Earth's surface and relay it back to a ground station. A sun-synchronous orbit, almost mirroring a CORONA mission profile, was kept at a 450-mile circular polar orbit to give 100 percent daily coverage of the Northern Hemisphere at latitudes above sixty degrees. The periodic passes over the central United States allowed direct downlink to Air Force Global Weather Central, near Omaha, Nebraska.

SUPPORTING THE NATIONAL RECONNAISSANCE PROGRAM

The early DMSP Block 1 satellites were mated to Scout rockets and launched from Point Arguello Launch Complex D (PALC-D), later designated Space Launch Complex 5 (SLC-5). SLC-5 launched five Block 1 satellites, with a dismal success rate. Out of the five launches, only two satellites reached orbit, with one placed into an improper orbit that hindered its primary mission. The one successful satellite, launched on August 23, 1962, was declared "end of mission" ten months later. To its credit, this satellite also played a crucial role during the Cuban Missile Crisis by providing weather data to SAC forces. This data helped reduce the number of airborne weather-reconnaissance sorties required over the Caribbean nation.

Most of the DMSP failures occurred due to various problems with the Scout booster. The NRO requested design changes, but since the rocket was purchased through NASA and the agency refused to acquiesce, DNRO Brockway McMillan cancelled two remaining booster orders and a follow-on contract. Hedging his bets, Colonel Haig had sought a launch vehicle replacement after the fourth Scout launch. Discovering a large number of excess Thor IRBMs recently returned to the United States from the United Kingdom, Haig pushed to convert these to space launch vehicles (SLVs) for the DMSP program. The upper stage to the Thor would rely on a solid-propellant rocket motor similar to the Scout fourth stage, allowing the same spin-table to release the satellite into orbit. Additional changes to the guidance system and attitude control section would characterize the "Thor/Burner I," as the rocket was designated. Months later, additional modifications to the upper stage, such as a solid-propellant, self-guided platform, would give rise to the "Thor/Burner II" configuration.

TABLE 4. THOR BURNER I LAUNCHES FROM SLC-10

Date	Pad	Launch Configuration	Launch Name	Program Association	Comments
19 Jan 65	4300 B-6	Thor-Burner I	Astral Lamp	DMSP	Payload failed to separate from second stage
18 Mar 65	4300 B-6	Thor-Burner I	Astral Body	DMSP	Successful
20 May 65	4300 B-6	Thor Burner I	Royal Eagle	DMSP	Successful
10 Sep 65	4300 B-6	Thor-Burner I	Victoria Cross	DMSP	Successful
8 Jan 66	4300 B-6	Thor-Burner I	Persian Lamb	DMSP	Second stage did not separate from Thor
31 Mar 66	4300 B-6	Thor-Burner I	Resort Hotel	DMSP	Last Thor-Altair (Burner I); successful

Before the first planned Thor/Burner I launch from SLC-10, the gap in weather coverage would be closed by a pair of Thor/Agena-D SLVs launched in early and mid-1964. DNRO McMillan approved transfer of two Thor/Agenas, the same launch vehicle for the CORONA reconnaissance satellites. Since these rockets were larger and more powerful than the previous Scout launch vehicles, two Block 1 satellites flew on each launch.[147] The two launches from SLC-1 placed four DMSP satellites in orbit successfully, with the last one surviving almost two years on orbit.

The next configuration of DMSP was designated Block 2. The changes on the satellite were in mass, due to additional sensors and the ability to provide direct readout to ground user terminals instead of direct downing to the STC in Sunnyvale or Global Weather Central (GWC) in Omaha, Nebraska:

> *These 160-pound vehicles, identical in size and shape to their 100-to-120 pound Block 1 predecessors, also mounted improved infrared radiometers and were known collectively as Block 2. Launched during 1965 and 1966, two of them attained Earth orbit and provided tactical meteorological data for operations in Southeast Asia. A fourth satellite, the one equipped and launched expressly for tactical uses on 20 May 1965, came to be called Block 3.*[148]

The "tactical users" were warfighting headquarters in Vietnam and aircraft carriers off the coast. The direct readout terminals provided imagery to tactical air mission planners, who needed up-to-the-minute information before the planes left the carriers' flight deck. Air-to-air refueling and rescue operations were two critical missions that used DMSP direct downlink data. The ability of tactical users to grab the satellite's data helped to remove the weather reconnaissance mission from the USAF's portfolio. The single Block 3 vehicle allowed the NRO to support the warfighter directly, as the other DMSP satellites continued to support the NRP. In the 1960s, this was a radical change from the NRO's directed missions but a welcomed departure. Years later, a program office in the Pentagon called TENCAP (Tactical Exploitation of National CAPabilities) would provide a similar approach of tactical users (e.g. soldiers on the ground) using national systems.

HEXAGON's "Buddy"

The enormous film-load on HEXAGON provided analysts with a very large intelligence take, provided that the images were clear and unobstructed. So critical was weather to the satellite's operation and presumed utility that it was considered key to HEXAGON's success:

> *Weather support comprised a continuous cycle during HEXAGON operations. Climatological data were used extensively during the mission planning stage to help in selecting the launching date and time and as an input to mission planning software that affected such factors as requirement weights (priorities), film allocation, weather thresholds and requirements satisfaction goals. Climatology also played an important role in on-orbit operations. For example, if the probability of successful coverage of South China was highest in December and January, marginal opportunities for photography could be passed up in August or September to concentrate collection efforts in the months with a higher probability of success.*[149]

In order to use DMSP data for various other programs, SAC ran the GWC. Full-day forecasting began with predictions from the early morning "scout" satellite, and verification was provided by the follow-on afternoon satellite:

> *Beginning in 1965 two DMSP polar orbiting, sun-synchronous weather satellites would normally function in circular orbits at 450 nautical miles altitude. One, a morning bird, passed over the Soviet Union about 0700 local time and relayed weather conditions at first light. A second, late morning (but called a "noon") bird began the same track about 1100 local time, showing the change in cloud cover with the increase in atmospheric heating during the day.*[150]

However, due to unexpected failures and optimization of the satellite constellation, having both scout and follow-on satellite passes was not always attained. Currency of data was also backed by collection of weather stations around the world, including a few located in the Soviet Union. This would give HEXAGON operators fresh data as "old" as two hours.

TABLE 5. THOR-BURNER II LAUNCHES FROM SLC-10

Date	Pad	Launch Configuration	Launch Name	Program Association	Comments
16 Sep 66	4300 B-6	Thor-Burner II	Irish Duke	DMSP	Successful
8 Feb 67	4300 B-6	Thor-Burner II	Arrow Point	DMSP	Successful
23 Aug 67	LE-6	Thor-Burner II	*	DMSP	Successful
11 Oct 67	LE-6	Thor-Burner II	*	DMSP	Successful
23 May 68	SLC-10W	Thor-Burner II	*	DMSP	Successful
23 Oct 68	SLC-10W	Thor-Burner II	*	DMSP	Successful
23 Jul 69	SLC-10W	Thor-Burner II	*	DMSP	Successful
11 Feb 70	SLC-10W	Thor-Burner II	*	DMSP	Successful
3 Sep 70	SLC-10W	Thor-Burner II	*	DMSP	Successful
17 Feb 71	SLC-10W	Thor-Burner II	*	DMSP	Successful

The accuracy of forecasts from GWC was impressive. Weather records note that two-thirds of Europe and Asia are covered with clouds in the midday hours on any random day. HEXAGON's success rate for cloud-free images hovered from 70 to 85 percent. On occasion, target photos deemed of high priority would be taken regardless of predicted or actual weather conditions.

Satellite operators inside the Satellite Test Center in Sunnyvale, California, factored in weather to the complex calculus for mission planning:

Cradle to Grave

Not only did DMSP's data assist in targeting and timing on the photographic satellites, but they also assisted in the recovery efforts. First-generation KEYHOLE satellites used cameras and film with recoverable reentry vehicles. These film "capsules" were retrieved either in midair or after a splashdown in the Pacific Ocean by specially modified aircraft or divers, respectively. The first KH-9 HEXAGON film capsule, however, sunk to the bottom of the ocean at a depth of 16,400 feet.[152] After a lengthy search, recovery efforts were hampered by extreme weather conditions, such as twenty-knot winds and two-foot seas with over six-foot swells, no doubt reported to the recovery ships, through the Fleet Weather Center, from DMSP data.

[T]*he complexity of the system—including the sensor subsystem—required that all control of the satellite be done by the Satellite* [Test] *Center (STC) at Sunnyvale, California. It was decided that the list of requirements (targets and target areas), with their priorities,* [would be sent] *to the STC where actual target selection for a particular revolution would be made (considering weather conditions and vehicle health) and sent as a command message to the satellite.*[151]

DMSP AND APOLLO 11

As word spread through unclassified channels about the usefulness of DMSP weather data, the urge to declassify the system and its information came to a fever pitch. One anecdote involving the returning astronauts from the first moon landing showed the utility of DMSP data, as well as the dangers in keeping it classified.

In a 2005 article for the Center for the Study of National Reconnaissance, senior NRO research historian Noel A. McCormack interviewed then-captain Hank Brandli about his assistance during the Apollo 11 mission in 1969:

> *The U.S. Air Force meteorologist had classified information indicating danger to the Apollo 11 crew returning to Earth from their historic mission. They had done it—the Eagle had landed. Neil Armstrong and Buzz Aldrin had walked on the moon, raised the American flag, collected samples, and then blasted off for a perfectly executed lunar orbit rendezvous with Michael Collins in the command module Columbia. Now they were headed home on the final leg of the trip for a July 24 splashdown in the Pacific*

Ocean. However, from his highly classified weather forecasting work, Capt. Brandli realized that instead of a heroes' welcome, the astronauts could face a watery grave.[153]

As the moon walkers were traveling back to Earth, weather data taken from the still-classified DMSP satellites showed the planned splashdown point in the Pacific Ocean would be covered with treacherous winds. These clouds, nicknamed "Screaming Eagles" due to formations looking similar to an image of an eagle's head facing west, contain rain showers and gusty surface winds up to twenty-five knots. The winds can amplify immediately and cause intense thunderstorms. While relatively unknown in the Atlantic (where most moon mission planning took place), Screaming Eagles were common in the Pacific Ocean from Hawaii to the equator. The danger with the winds was primarily against the command capsule's parachutes; the winds generated could have shredded the chutes, causing the capsule to drop into the ocean.

Brandli was a meteorologist at Hickam AFB, working as a weather prediction specialist with data from Program 417 (DMSP Block 4A and 4B) satellites. As recounted previously, aside from predicting cloud cover for NRP satellites, DMSP data was critical for snatching the returning film capsules in midair. While Brandli had used sanitized data during a tour in Vietnam, it wasn't until he was assigned to Hickam that he learned the true nature of DMSP's mission: "I say [*sic*], Holy Smokes, that's what this weather satellite is for—to support Corona! We wanted the best weather information so we could turn the cameras on over the Soviet Union and China and Cuba."

His support of the 6594th Test Group and their job of grabbing returning CORONA, GAMBIT and HEXAGON capsules was even compartmentalized within his chain of command. Even his vice-commander, second in charge, was not briefed to the true mission.[154]

While not just "any" Air Force meteologist, Brandli was a recognized pioneer in the field of satellite meteorology. He discovered that by using the high-resolution data, weather could be forcasted up to five days in advance for an area from the equator up to twenty-five degrees of latitude. Within the final seventy-two hours of the Apollo 11 mission, Brandli had a problem. He had seen the right mix of weather to create Screaming Eagles over the Apollo 11's landing zone in the Pacific Ocean. But because he was one of a handful briefed to the DMSP mission, he could not warn the planners at NASA.

Taking the initiative to find someone "in the know" (cleared to BYEMAN), Brandli was informed that the U.S. Navy was providing the forecasting data to NASA for the recovery mission, so he reached out to the chief weather officer at the Fleet Weather Center in Pearl Harbor: Navy captain Willard "Sam" Houston Jr. Ironically, Captain Houston had also been briefed on Program 417, so Brandli brought him in to the 6594th Headquarters building to show the photos. While the proof was incontrovertible, the burden in passing along the critical classified data was placed into Captain Houston's lap.

Rear Admiral Donald C. Davis took Houston's advice that the landing site needed to be altered from the coming storms without revealing the photos behind it. Davis rerouted the USS *Hornet* and its carrier task force to the new landing coordinates, and the astronauts were successfully retrieved.

It wasn't until 1995, when the CORONA program was declassified,[155] that Brandli was able to reveal the details behind the Navy Commendation medal he received from Chief of Naval Operations Elmo R. Zumwalt Jr. for his assistance in the Apollo 11 recovery. Brandli spoke about the incident: "It was a huge undertaking to move the carrier recovery fleet and convince the 'powers that be' to change the landing site. Captain Houston did a hell of a job. I often wonder: if it had been anyone else, would it have happened the same way?"

Discussing the event after CORONA's declassification, Houston revealed to Brandli: "They sent reconnaissance aircraft out to check [the weather], and we were right on the money."[156]

Soon after the Apollo 11 recovery, two senior officials—Dr. John McLucas and Dr. Robert Naka, director and deputy director of the NRO, respectively—visited the 6594th Test Group in Hawaii. Not knowing about their connction to the NRP or the NRO, Brandli showed both men the Screaming Eagle photos and other unique images from the DMSP satellites.[157]

END OF DMSP CLASSIFICATION

Competing funding for both a military and civil weather satellite system caused policymakers to try to merge the two programs or find more utility from the data obtained. NRO director John McLucas wrote in a memo to Secretary of Defense Melvin Laird:

Technicians inspect a DMSP weather satellite inside SLC-10W. *Courtesy of Vandenberg Space and Missile Technology Center.*

The [Defense Meteorological Support Program], *formerly Program 417, is another example. This program was started in August 1961 by the NRO to provide weather observations over the Sino-Soviet Bloc for our photographic reconnaissance satellites. Over the years the program has been successively removed from BYEMAN Controls, came under SPECIAL ACCESS REQUIRED Controls and this year was released from that control but retained a SECRET classification. It is vital to our successful operations.*

[DMSP image] *resolution is not sensitive from a reconnaissance standpoint, but is of higher quality than the weather satellite pictures released to the public from the national weather system. The* [DMSP] *photography has in the past been handled as SECRET information.*

This year I initiated discussions with the Secretary of Commerce with the desire to join forces to make maximum use of the weather information from both the [DMSP] *and the programs managed by the National Oceanic and Atmospheric Administration (NOAA). We have proceeded remarkably well toward this goal, and have arranged to furnish NOAA* [DMSP] *data on an unclassified basis. This action requires that we be prepared*

> *to acknowledge the existence of a Department of Defense space system which provides meteorological data—but we will not, however, release any operational information. I regard this as a very significant achievement in best use of resources in the national interest.*[158]

To the USAF chief of staff, McLucas wrote:

> *As you are aware, I have been appointed by Secretary Rush to work out methods of improved cooperation with NOAA in using data generated by weather satellites. At present NOAA receives little benefit from our own classified meteorological satellite system* [DMSP]. *Before we can provide NOAA routine support from the* [DMSP] *we must declassify the data and alter our security policy.*[159]

Upon the launch of DMSP in November 1972, all DMSP data would be unclassified except for sensitive specialized tasking that would reveal military operations.

> *The primary global information is two nautical mile resolution visual and infrared data which will be available to NOAA, unclassified, through the Air Force Global Weather Central. The primary tactical information is high resolution (one-third nautical mile) visual and infrared data. Mainly intended for direct readout, a limited amount of high resolution data are stored on tape recorders for subsequent transmission to the Air Force Global Weather Central. High resolution data over the continental U.S. and adjacent waters would be routinely available to NOAA, unclassified, as well, as other regions on an as-requested, as-available basis (e.g., tracking of tropical storms which are a potential hazard to the US). In addition, automated analyses using all available weather data are available to NOAA. Technical details of the* [DMSP] *spacecraft would remain SECRET.*

Declassification and announcing the declassification are two total separate events in the eyes of the NRO and Intelligence Community.

> *1. The declassification and release of* [DMSP] *weather products to the new* [sic] *media have posed a difficult problem for SAMSO, primarily because the SAMSO/*[DMSP] *involvement in the program remains classified.*

2. We anticipate questions about the satellite directly from the media and during public appearances by SAMSO representatives. If our involvement in the weather satellite program can be declassified, then SAMSO can respond affirmatively, without embarrassing the Air Force or violating security directives.

3. It appears that the best interest of the DoD and the Air Force can be served by declassifying SAMSO involvement, reviewing other elements of the program for declassification, and initiating a low key information policy; i.e. no voluntary release, but prepare a list of answers to legitimate queries.

All DMSP satellites since Block 4 were provided direct readout capability to tactical users worldwide. Weather terminals were installed on aircraft carriers, at ground stations and at select bases around the world. The data would be analyzed by meterorologists and assist in air operations planning efforts. When the Block 5 series was developed, the data was provided in an encrypted format, lessening the chance of interception by authorized users.

TABLE 6. THOR-BURNER IIA LAUNCHES FROM SLC-10

Date	Pad	Launch Configuration	Launch Name	Program Association	Comments
14 Oct 71	SLC-10W	Thor-Burner IIA	*	DMSP	Successful
24 Mar 72	SLC-10W	Thor-Burner IIA	*	DMSP	Successful
9 Nov 72	SLC-10W	Thor-Burner IIA	*	DMSP	Successful
17 Aug 73	SLC-10W	Thor-Burner IIA	*	DMSP	Successful
16 Mar 74	SLC-10W	Thor-Burner IIA	*	DMSP	Successful
9 Aug 74	SLC-10W	Thor-Burner IIA	*	DMSP	Successful
24 May 75	SLC-10W	Thor-Burner IIA	*	DMSP	Successful
19 Feb 76	SLC-10W	Thor-Burner IIA	*	DMSP	Failed to reach operational orbit
11 Sep 76	SLC-10W	Thor-Burner IIA	*	DMSP	Successful
1 May 78	SLC-10W	Thor-Burner IIA	*	DMSP	Successful
6 Jun 79	SLC-10W	Thor-Burner IIA	*	DMSP	Successful
15 Jul 80	SLC-10W	Thor-Burner IIA	*	DMSP	Failure

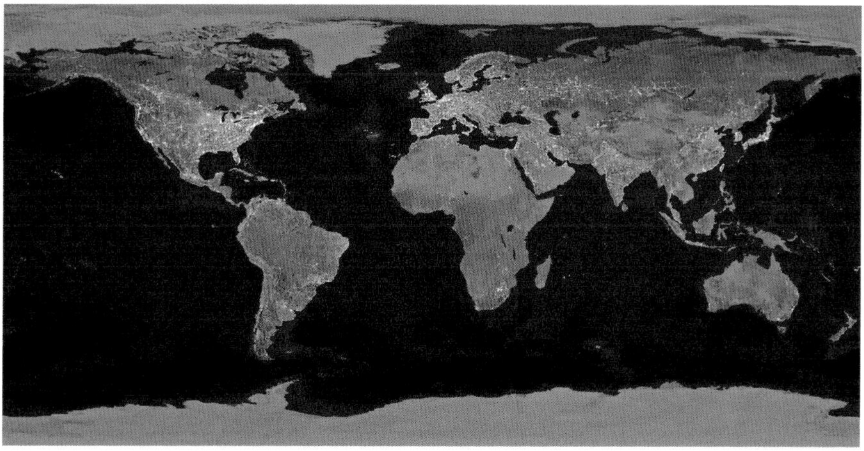

This nighttime mosaic of the world was pieced together from DMSP data. The U.S. Eastern Seaboard is rife with illumination, whereas Central Africa, Siberia and North Korea are virtually blacked out. *Courtesy of NASA.*

At the time of the final launch, DMSP plans directed the continued use of upgraded versions of the Block 5D spacecraft indefinitely. When Thor was eliminated from the service's inventory, the Air Force directed the transition of DMSP to the Atlas booster in 1982. More sensors equaled more spacecraft mass, and improvements in encryption and survivability necessitated the move to the larger booster. Since the reason for the DMSP program was primarily cloud cover detection, no major changes to the cloud cover imager were provided to the 5D spacecraft.

DMSP's Legacy

Like its Thor launcher, the DMSP program provided many great accomplishments to astronautics and satellite programs. The first contribution was to space management techniques. In the early days of space launches, programs were underfunded, lacked clear direction and held an uncertain future (sometimes due to the funding and lack of focus). Lieutenant Colonel Haig gave a laser-sharp focus to the "small" program with no outside support contractor to run roughshod over engineering decisions. Mirroring the earliest CORONA effort, quick decisions were made, and work was accomplished just as rapidly. The second contribution was engineering genius—the use of Earth's magnetic field to keep the rotating satellite spinning. Since the earliest DMSP blocks used a spinning satellite to maintain its orientation toward Earth. DMSP engineers realized that instead of using chemical rocket thrusters, which would eventually run out, electrical torquers could be used to keep the satellite spinning at a constant rate. Contributions were focused on the ground stations supporting DMSP. The tracking techniques developed for DMSP were used for other satellite programs after an Air Force captain prepared a less complex (and cheaper) way of tracking the low-flying satellites as they whipped across the sky quickly.

> *Written into RCA's second-year contract for an additional four Block 1 satellites, the concept worked in space. Third, when the DMSP ground stations were assembled in 1963, the program office eliminated the costly "boresight tower" used routinely to determine a tracking/readout antenna's pointing vector and a transmitter used to check the receiving system sensitivity during operation. Program personnel substituted instead a technique of scanning the sun to establish the pointing vector with a hermetically sealed low-energy transmitter in the center of the antenna reflector used to check receiving sensitivity. The DMSP station test procedures worked just as accurately and at far less cost; they became standard practice for nearly all tracking/readout systems.*[160]

The progam has continued into the twenty-first century, albeit no longer launched on Thor SLVs. The heavier satellites were launched on Atlas and Titan II space launch boosters. One booster in 2006 was launched aboard a Delta IV, renewing the legacy of DMSP and Thor-related hardware. Testimony from Undersecretary of the Air Force Peter B. Teets in 2004 succinctly cemented the legacy of DMSP in the space warfighter's toolbox, while simultaneously putting a nail in the program's coffin:

> *Recent conflicts have proven, once again, how vital meteorological forecasting is for military operations. Knowing what the weather is in any given location allows us to choose the right weapon for the right target, and is an invaluable asset for navigation. The National Polar-orbiting Operational Environmental Satellite System (NPOESS) will satisfy both civil and military national security requirements for space-based, remotely sensed environmental data that will significantly improve weather forecasting and climate prediction. NPOESS is a tri-agency (DoD/Department of Commerce* [DOC]*/NASA) satellite*

A model of a DMSP Block 5D weather satellite (minus the solar panel) sits inside the museum at SLC-10. *Courtesy of the author.*

DMSP Today

The DMSP mission is to collect and disseminate global, high-resolution visible and thermal cloud cover imagery and other critical air, land, sea and space environment data to Department of Defense forces and the intelligence community. DMSP data also is furnished to the civilian community through the Department of Commerce. The current DMSP satellites are managed by the Remote Sensing Systems Directorate at the U.S. Air Force Space and Missile Systems Center. Command and control of the DMSP constellation are provided by the National Oceanic and Atmospheric Administration Satellite Operational Facility in Suitland, Maryland, with a backup at the 6th Space Operations Squadron (6 SOPS), Schriever AFB, Colorado.

program consolidating the missions and programs of DoD's Defense Meteorological Satellite Program (DMSP) and DOC's Polar-orbiting Operational Environmental Satellite (POES) systems into a single integrated program. An integrated suite of 12 very complex instruments will provide visible and infrared cloud-cover imagery and other atmospheric, oceanographic, terrestrial, and space environmental information. The system is currently in development, with a planned first launch in FY10.[161]

Even with the proposed date given to Congress, the planned merger of DMSP and POES into NPOESS never occurred. The political merging of the programs forced the DMSP assembly line to cease in preparation for future NPOESS satellites. Failure of the NPOESS program, along with its planned successor, the Defense Weather Satellite System (DWSS), left the Intelligence Community and U.S. forces with—you guessed it—DMSP. As a matter of pride for all involved, the legacy of success for DMSP was solidified with the launches at SLC-10.

Chapter 7

NEW LEASE ON LIFE

Changes in national launch policy after the Apollo moon program had an adverse effect on the Thor program. Recommendations from the Space Task Group (STG) formed by President Nixon in 1969 highlighted the need for a "space truck" to ferry payloads to orbit. The solution was to be cost effective for both the civil and military space programs—a one-size-fits-all solution. The STG came back to Nixon in early 1969, before the first moon landing, and recommended a reusable space plane; colloquially called a "space shuttle," the official name would be Space Transportation System (STS). The Air Force had done studies on reusable spacecraft, the DYNA-SOAR for example, but wanted mature technology that could be made operational, lessening the then-current burden of too much intervention to monitor and maintain the orbital systems. NASA, riding high from the success of the Apollo program, desired a capability to keep humans in space with regular spaceflight. The DOD, and its hidden courtier of the NRO, had the budget, and NASA had the need. A compromise was reached, albeit ill fitting for all parties: "No major launch-vehicle development was under consideration in the Air Force. Although not expecting the shuttle to replace all current launch vehicles, the tactic was to push NASA to provide as much capacity as possible. Whatever was left over would have to be accommodated by expendable boosters."[162]

THE FINAL LAUNCH

On July 14, 1980, the final launch at SLC-10W took place with a twenty-one-year-old Thor 301 first-stage.[163] The rocket's liftoff was within standard flight parameters. Almost one-half second later, however, the Thor's telemetry signal ceased. During the early stages of flight, the rocket appeared to follow its orbital profile. During the separation of the payload, the telemetry signal of the satellite dropped out. Tracking assets belonging to the Western Test Range, such as the Advanced Range Instrumentation Aircraft, started a wideband search for the signal through the electromagnetic spectrum, with no luck. After twenty-six minutes and thirty-three seconds, the satellite was declared lost. It was a disappointing end to the SLC-10 launch legacy.

An investigation board reviewed the processing of the DMSP and launch of the Thor booster and released its findings in August 1980. The three recommendations of the investigation came too late for the Thor program:

1) The contractor should revamp its procedures for inspection, rework and retest of discrepant propulsion hardware
2) An independent team should review the requirements, engineering procedures and test program of the Block 5D-2 reaction control equipment (RCE); and
3) A test should be conducted in vacuum simulating a hydrazine leak.[164]

The next DMSP spacecraft, Block 5D-2, had a 520-pound increase in weight from the previous version. This necessitated the use of a larger booster, the Atlas E/F space launch vehicle. These refurbished Atlas ICBMs were plentiful from the ICBM force modernization and missile base closures in the 1960s and were a ready solution for DMSP launches. By the final launch in 1980, the Thor inventory was near depleted, and the DMSP block changes took away Thor's final customer. While the move to the space shuttle had been simmering in the background for the entirety of the 1970s, the final death-knell to the Thor program came with the publishing of President Ronald Reagan's National Security Decision Directive 8, *Space Transportation System*, on November 13, 1981:

> *The United States will continue to develop the STS through the National Aeronautics and Space Administration in cooperation with the Department of Defense to service all authorized space users. The STS will be the primary space launch system for both United States military*

> *and civil government missions. The transition to the Shuttle should occur as expeditiously as practical.*[165]

While the Thor program officially ended in July 1981, its legacy continued.[166] The Extended Long Tank Thor first stages were built inside McDonnell Douglas's Huntington Beach facility for its Delta II medium-lift rockets. After the space shuttle Challenger exploded in 1986, the U.S. government reversed its decision to place all national security payloads on board. Rocket assembly lines were restarted, and Thor variants such as the Delta II and Delta III continued on, combining tried-and-true Thor hardware with present-day technological modifications.

Commentary on the End of Thor

Interesting commentary on the end of the Thor program was written by Lieutenant Colonel Wayne Eleazer, USAF (retired), on the Space Review website. Serving as the last Thor program manager in the Air Force, Lieutenant Colonel Eleazer stated: "[The Air Force] had nine Thor boosters still in storage, plus a few bits and pieces. The rockets could still be made to fly just fine but it was politically incorrect to even mention the fact….[A] call from NASA consisted of a stern lecture reminding the Air Force that the space shuttle was the officially anointed launch vehicle and NASA would brook no competition. Never mind that the boosters were sitting in a warehouse; they could not be used."[167]

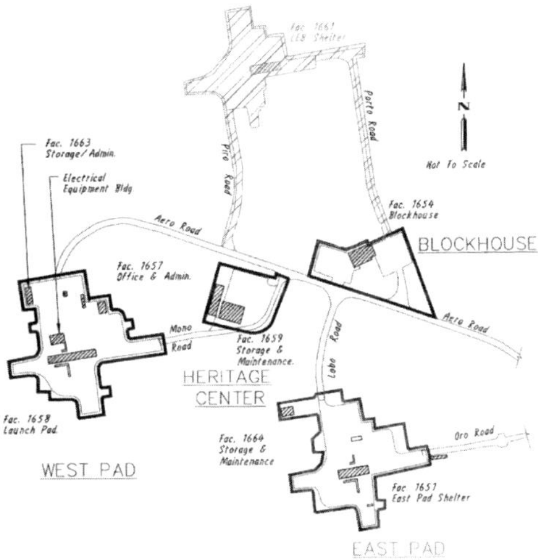

The National Historic Landmark boundaries of SLC-10 include the East and West Pads, Heritage Center buildings and the Blockhouse. LE-8 was not included in the NHL designation. *Courtesy of the author.*

NATIONAL HISTORIC LANDMARK

Unlike the gala events surrounding the last Titan launch in 2005, the ending of the Thor program in 1981 was neither celebrated nor largely acknowledged by the Air Force. The site would sit dormant for three years, with the invasive ice plant *Carpobrotus edulis* growing across the complex. Temporary respite arrived in 1984, when the National Park Service (NPS) initiated the "Man in Space" National Historic Landmark (NHL) theme study. Vandenberg AFB was one of many nationwide military installations investigated for their roles in aerospace history. At first, NPS researchers selected SLC-2 for the NHL designation since it was the site of the first Thor IRBM launch. Additional research showed the modifications that SLC-2 endured covered a good portion of the Thor family legacy but altered the facilities beyond the scope of the NPS investigation. Over its history, SLC-2 launched the following variants of the Thor IRBM:[168]

- Royal Air Force (RAF) Thor IRBM
- Thor Agena B
- Thor Agena D
- Thor Able Star
- Thrust Augmented Thor (TAT) Agena D
- Thrust Augmented Thor (TAT) Delta
- Long Tank Thor (Thorad) Agena D
- Long Tank Thor (Thorad) Delta
- Delta
- Delta II

In SLC-2's place, SLC-10 was put forward as an alternative site for the "Man in Space" study, as it supported missile combat training launches until 1962 and booster launches until 1980. In comparison, SLC-10 hosted four variants of Thor:

- Royal Air Force (RAF) Thor IRBM
- Thor Burner I/Altair
- Thor Burner II
- Thor Burner IIA

After the dismantling of the launch shelters after the last Thor IRBM launch (1959), the pilfering of equipment for Operation DOMINIC (1962)

Left: Major General Donald Aldridge, 1st Strategic Aerospace Division commander, delivers remarks during the SLC-10 National Historic Landmark dedication on July 18, 1986. *Courtesy of Vandenberg Space and Missile Technology Center.*

Below: The National Historic Landmark plaque at SLC-10 sits outside the Admin building on the walk toward the West Pad. *Courtesy of the author.*

and then the rebuilding of the pads with returning EMILY hardware (1963), SLC-10 existed as "the best surviving example of a launch complex built in the 1950s at the beginning of the American effort to explore space."[169]

On July 18, 1986, the commander of the 1st Strategic Aerospace Division, Major General Donald O. Aldridge, presided over the ceremony to designate SLC-10 as a National Historic Landmark.[170] After his remarks on the West Pad, General Aldridge and the Department of Interior representative unveiled the commemorative plaque denoting NHL status. Even after the ceremony, the site remained unused for many years. In 1987, the Air Force gave SLC-10 responsibilities over to the 4315th Combat Crew Training Squadron for restoration and conversion into the Vandenberg Space and Missile Heritage Center.

PART II

THE PRESENT

Chapter 8

VANDENBERG SPACE AND MISSILE HERITAGE CENTER

Historic artifacts inside the Heritage Center show how operations at SLC-10 contributed to the intermediate-range ballistic missile training program, the antisatellite efforts of Program 437 and the launches of the DMSP satellites. Additionally, hardware from the other programs are presented to interpret the evolution of missile and spacelift activity through military, commercial and scientific missions at Vandenberg from the beginning of the Cold War through the present day.

Exhibits listed inside the center include:

- A full-sized launch console from an Atlas ICBM launch control center
- A scale model of a Titan II Missile complex
- Mark 21 reentry vehicle mock-ups from the now-retired LGM-118A Peacekeeper missile
- Display of Peacekeeper test flight profile
- One-third scale model of DMSP Block-5D satellite
- Minuteman II Commander's Launch Console (complete with working lights!)
- Minuteman II Deputy Commander's Console
- Thor rocket engine
- Agena upper stage
- Mark II reentry vehicle
- A complete (albeit disassembled) Minuteman II ICBM

A preserved Atlas-D console is one of the centerpieces of the museum collection, displaying artifacts important to the history of space launch at Vandenberg AFB. *Courtesy of the author.*

- Titan ICBM first-stage engines
- A replica of a Minuteman Launch Control Center Blast Door
- Photos of the CORONA program
- Personal items belonging to Brigadier General William King
- Scale model of SLC-4E complete with Titan IV replica
- Mission flight control console from the Western Range
- Program 437AP model

ACTIVITIES AT SLC-10

While its distant location from Vandenberg's main garrison usually precludes numerous visitors during a normal week, SLC-10 hosts a number of activities throughout the year. Educational, social and ceremonial events occur at various locations around SLC-10. During the leadup to its scientific launches, NASA hosts a "NASA Social" event that links influential bloggers, writers and tech geeks with tours to see and experience what the civilian space program has to offer all Americans. For example, in 2014 the NASA social cadre came out in preparation for the Orbiting Carbon Observatory-2 (OCO-2) launch. The cadre was treated to behind-the-scenes tours of NASA facilities at Vandenberg, briefings held by agency administrators and a tour of SLC-10. The tour of the complex was relevant due to OCO-2's Delta II launch booster being a "grandchild" of the Thor IRBM. While many of the SLC-2 Delta II facilities are unique (e.g. the gantry and processing buildings), many legacy Thor facilities and hardware are still in use by the contractors.

Many students, local and international, have toured the facility and received special attention on science and mathematics in terms of rocket science. Children from as far away as Israel and Russia have ventured to SLC-10 to learn about its history and perform hands-on STEM activities. Most agencies and contractor companies on Vandenberg do not celebrate the annual "Bring Your Kid to Work Day"; for those that do celebrate it, a trip out to SLC-10 offers a peek into what some parents do for a living. After the children see the models and historic pictures, they build (and test) their own paper rockets with a compressed air launcher.

Military members tour the inside of the West Pad shelter. A tour at SLC-10 is part of the initial space training curriculum for junior officers and enlisted members. *Courtesy of the author.*

Visitors to SLC-10 listen as SAMTEC director Jay Prichard explains the heritage of the space and missile artifacts surrounding them and why, after decades, their story is still relevant. *Courtesy of the author.*

Members of the Joint Space Operation Center (JSpOC) gather to celebrate at SLC-10 during a Museum Barbeque (Mu-Q) in February 2015. *Courtesy of the author.*

Personalized tours for special gatherings, such as the NRO cadre at Vandenberg, allow personnel a chance to see how their predecessors met the challenges of the past and how present-day operations are still vitally important.

The museum location draws a lot of visitors to SLC-10, but the connection to missiles and space of days gone has many uniformed personnel returning to the site for personal reasons. When a military member is recognized with potential to lead at a higher rank, a promotion ceremony is held to formally announce the progression of rank and responsibilities. Promotion ceremonies have been held inside the West Pad shelter, atop the Blockhouse and, for those missileers and spacelift officers wanting a historical connection, out on the pad at LE-8. How many missile combat crew members can boast of receiving their missile badge on the same spot as the first operational missile launch? Or how many newly certified space operators can stand on the grounds of the first Blue Suit space launch organization in the USAF and receive their hard earned space operations wings? One enterprising young staff sergeant from the Vandenberg AFB Civil Engineering squadron reenlisted in the attic of the Admin building; she had worked diligently on removing and reinstalling the electrical system during the restoration of the Admin building in the late 1990s. By taking her reenlistment ceremony into the crawl spaces above, she showed her collegues and supervisor how she "owned" her work. Even when the work was hidden inside the walls and attic, the mantra of "owning it" reflects the best traditions of honorable service like the work performed at SLC-10.

Retirement ceremonies and going-away gatherings are a bittersweet occasion, recognizing an individual or family's sacrifice for the good of the nation. The gatherings have ranged from the serious (i.e., catered dinner) to the jovial, such as a good ole "Mu-Q" (Museum Barbeque) behind the Admin building. The father of the Intercontinental Ballistic Missile (ICBM) program, General Bernard Schriever, celebrated a birthday out on SLC-10's West Pad.

One of the activities performed regularly at SLC-10 is a team-building exercise, part scavenger hunt and part history knowledge quiz. The military prides itself on creating cohesion within groups, allowing these activities as an integral part of leadership development. During the execution of these events, many high-ranking officers have been stumped by the quiz (even when profusely arguing about the "actual answers" from their own experiences). However, due diligence to verify the availability of quiz answers within the exhibits was performed on more than one occasion. The executive director used third-party individuals to validate the questions—usually someone with no space/missile experience or any preconceived knowledge of SLC-10 history. Included below is the text of the situation brief given to the small teams during the beginning of the scavenger hunt; however, to maintain the activity's fun and usefulness, the questions and answers are *not* provided:

Situation Report
18 SEP 2011: USSTRATCOM confirms USAF DMSP spacecraft to deorbit and impact the West Coast of the United States.

4 OCT 2011: Impact prediction from join space operations center states re-entry on Vandenberg AFB approximately mid-month.

Intel Brief
Recent USSTRATCOM developments have increased US concern regarding intelligence data collected through national technical means and partner nations. Foreign governments continue to show extreme interest in acquiring older (US) spacecraft sensor capabilities. It has been routine practice for Russian engineers to acquire and diligently reverse-engineer foreign hardware to be adapted for their particular mission needs.

Particularly disconcerting is the interest in thermal, magnetometric, optical interpolation and microwave technologies, all of which have been employed on even the oldest general of DMSP spacecraft. It is believed that with this hardware and technology, the Russian scientists intend to combine the capabilities with other commercially available products to narrow the accuracy window for a new anti-missile defense targeting system, potentially negating the effectiveness of the ground-based mid course interceptor system.

It is now believed Russian Spetsnaz (special forces) advanced reconnaissance units are mobilizing to seize any recoverable debris from downed US satellites. These teams travel in small, extremely mobile units and are known to carry various weapons to engage targets at multiple ranges. These advanced operators travel in pairs via motorcycle for rapid ingress and egress.

Recently observed "noise" overlapping base radio communication frequencies have been suspected to be data downloaded directly from Molniya Elint satellites to Spetsnaz teams about satellite impact points and debris recovery efforts.

The collected information seems to be finding aids, using unrelated riddles from a popular outdoor pastime known as geocaching. These riddles contain coordinates to the downed hardware.

Intel Update
It is now confirmed Spetsnaz forces are operating within Vandenberg's confines. Primary cell activity is centralized somewhere due west of the runway, possibly in the vicinity of Space Launch Complex 10. Spetsnaz teams are expected to convene at this location and deploy recover forces before egressing.

Mission Brief
Establish 4–5 person teams to recover the sensitive hardware before it can be secured by the adversary.

Engage the Spetsnaz cell at SLC-10. Once arriving at the Heritage Center, the ground assault team will be required to find the DMSP hardware and retrieve the intelligence information packet without detection by the Russian sentry and then run on foot to the designated course marker before returning to meet the logistics team. This team will be required to overpower hostile forces by demonstrating superior academic acumen in the primary exhibit areas.

Upon completion of these two tasks, each team may embark on search and recovery efforts. Each team will rendezvous at the Pacific Coast Club for mission debrief at 1600 hours.

Teams will be scored on the number of locations visisted and the items recovered. Failure to secure the required items will be considered a detriment to U.S. national security and will result in points lost.

Intel Update
The Spetsnaz mission plan has been intercepted through COMINT channels. An electronic copy will be made available to each team. This plan is presumed to include the actual locations of each item including directions and descriptions. Unfortunately, this file is protected by an enciphered password. The solution to this password cipher can only be acquired by defeating the Spetsnaz cell (through successful completion of the challenges at the Heritage Center). It will be up to each team individually to decode the cipher to acquire the password.

End Communique

PART III

THE FUTURE

Chapter 9

SAMTEC AND THE THOR HISTORIC DISTRICT

Preserving the Past, Inspiring the Future

In early 2015, the executive director of the Space and Missile Heritage Center began merging a vision of past, present and future at SLC-10: the Space and Missile Technology Center (SAMTEC).[171] While the current incarnation of SAMTEC is just a shadow of the future vision, it lays the foundation of preserving the story of the early days of missile activities and Vandenberg AFB through direct engagement with the public through STEM activities.

The SAMTEC name was chosen as a blending of the previous Vandenberg Space and Missile Heritage Center moniker while recognizing that direct application of history's lessons learned through STEM activities. The SAMTEC name has historic roots at Vandenberg. During the build-up of missile infrastructure at Vandenberg, ARDC, through AFBMD, supported SAC operations of launching missiles. This support multiplied over the next two years, enabling AFBMD to enlarge its field office into a wing organization. After two redesignations, the 6595th Aerospace Test Wing was aligned under the Space Systems Division in November 1961. Creation of the Air Force Western Test Range in 1964 and increasing numbers of space and missile launches brought the actions of the 6595th to the forefront of Air Force leadership. On April 1, 1970, the 6595th and AFWTR were realigned under the Space and Missile Test Center, SAMTEC. The name did not last beyond the end of the decade, with

Two Thor veterans, one of Project EMILY and the other of DMSP launch operations, pose in front of the booster inside the West Pad. *Courtesy of Vandenberg Space and Missile Technology Center.*

HQ SAMTEC being renamed the Space and Missile Test Organization (SAMTO) on October 1, 1979.

The reincarnation of the SAMTEC name came in 2015, when executive director Jay Prichard realized expansion of the center's mission to include hosting STEM activities would help drive home lessons learned by the original Cold War "steely-eyed missilemen":

> *My job is to kind of be the Indiana Jones of the Cold War with one hat, and with another hat, be Bill Nye the science guy....I find the story, preserve the culture that tells the story and then interpret that story to various audiences, whether they are a group of school kids, retired military members, or young Airmen going through training.*[172]

This captivating story can also be told with hands-on activities at SLC-10. The restoration of the Blockhouse, including the upstairs penthouse, will open up opportunities for hands-on learning and instruction.

Current STEM activities at SAMTEC include:

SAMTEC director Jay Prichard talks to one of the restoration technicians inside the East Blockhouse. *Courtesy of the author.*

- An amateur space tracking site, using publicly available data from www.space-track.org
- Astronomy and astrophotography setup to view predicted solar events, such as sunspots and eclipses
- Experiments with magnetospheric observations to monitor solar wind flux
- Launches with constructed paper rockets (soaring to over 150 feet!)

Planned additions for STEM activities include an initiative to get a deactivated telescope from the defunct Anderson Peak Observatory. The telescope is actually five instruments on a sturdy platform that assisted in data collection during Vandenberg launches. The scope tracked space/missile launches from over 120 miles away from the launch site. When the rocket/missile appeared above the horizon, the telescope tracked them with a slewing rate of fifteen degrees per second.

While images from the Anderson Peak telescope are amazing in their own right, additions to the communication infrastructure (e.g. fiber optics) would allow a real-time/near-real-time dissemination of images to visitors, similar

SAMTEC director Donald "Jay" Prichard inspects the interior of the Blockhouse during its 2014–15 restoration. Prichard has directed the heritage program at SLC-10 since the early 1990s. *Courtesy of the author.*

to other popular exhibts at many space and astronomy museums. Space and terrestrial weather reports directly downlinked from NOAA satellites or from online websites would provide students and visitors the ability to see how satellites, some of which were launched from Vandenberg, have assisted in everyday life.

The possibilities of linking STEM activites with satellite technology launched from Vandenberg are endless. One example is the Global Positioning System (GPS). GPS is the de facto standard for position, navigation and timing (PNT) information for most of the world. The first block of GPS satellites was launched from the Western Range aboard Atlas-E/F SLV boosters in the late 1970s/early 1980s. While the original NAVSTAR GPS command and control ground station at Vandenberg has been demolished, the legacy of the system's early days are preserved by SAMTEC. The first amateur radio satellite, OSCAR-1 (Orbiting Satellite Carrying Amateur Radio), was piggybacked aboard a Discoverer/CORONA launch in 1961.[173]

Thor Historic District

When the last Delta II rocket heads to space later this decade, an amazing story for the ages will come to an end. SLC-2 West, the last active launch site from the original Thor Complex 75, will finally be put to rest. In an April 8, 2015 article, *Spaceflight Now* reported that the final Delta II vehicle had not been allocated for any missions. The Delta II retirement is due to the lack of suitable payloads in the booster's weight class. The combined history of the Thor launch sites at Vandenberg hold their own pieces of the puzzle. SLC-2 and the surrounding sites of SLC-1 and SLC-10 all launched missile or space booster variants of Thor.

Aside from the National Historic Landmark (NHL) status of SLC-10, the other sites on Vandenberg have been recognized as having significant historical value. There are no current plans in the Air Force budget to preserve any of the Thor sites, beyond what is already at SLC-10. SLC-1's launch pads are in disrepair, with only skeletal remains of the environment shelter still standing on the East Pad. The launch pad at SLC-2E was decommissioned in 1972 and stripped of equipment and other salvageable materials. Only SLC-2W and SLC-10 survive.

In 1984, the NPS decided to concentrate preservation efforts on SLC-10 since the other sites had been radically modified from their original

Above: The Thor booster inside the West Pad shelter allows visitors to experience the environment from which the rocket was prepared and launched. *Courtesy of the author.*

Below: A Program 437 crew poses inside the East Blockhouse prior to training operations. *Courtesy of Vandenberg Space and Missile Technology Center.*

construction. Now, thirty years after the dedication of SLC-10, it is time for the Air Force and NPS to reassess the NHL criteria to encompass the entire Thor/Delta complex. The closeout of Delta II operations at SLC-2 allows another narrative to be preseved. The National Register of Historic Places (NRHP) recognizes significant properties, classified as buildings, sites, districts, structures or objects. In cases such as this, the NRHP also recognizes districts, sites that "possess a significant concentration, linkage, or continuity of sites, buildings, structures, or objects united historically or aesthetically by plan or physical development." Additionally, a district "derives its importance from being a unified entity, even though it is often composed of a wide variety of resources. The identity of a district results from the interrelationship of its resources, which can convey a visual sense of the overall historic environment or be an arrangement of historically or functionally related properties." A Thor Historic District at Vandenberg would conceivably preserve the sites and equipment from the earliest Thor missions to the present-day Delta II.

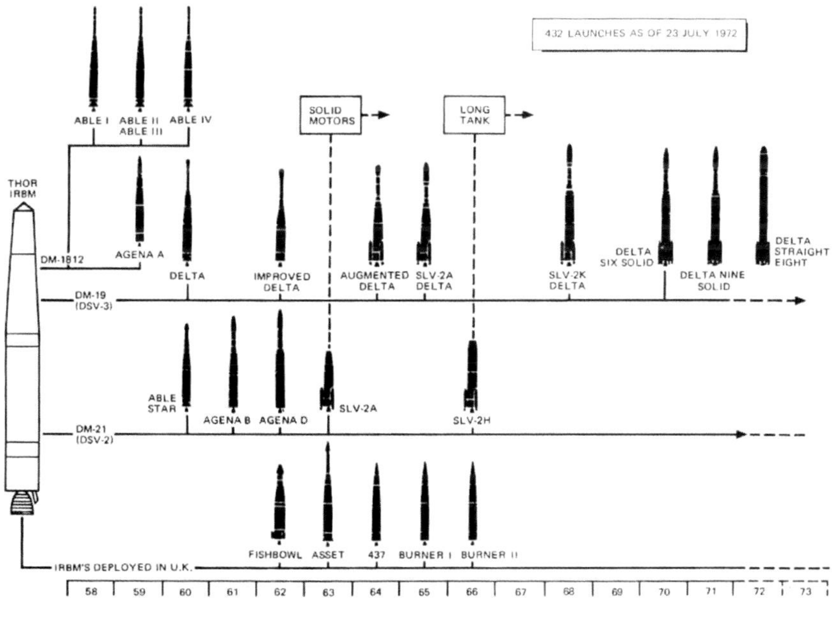

Thor-Delta Family Tree 1957-1973

This chart shows the lineage of the Thor IRBM until the early 1970s. The rocket and its successors would fly until the latter half of the 2010s. Note the bottom row for the fate of recycled EMILY boosters. *Courtesy of the United States Air Force.*

The final Delta II rocket is scheduled to launch from Vandenberg's SLC-2 in 2018. As of this book's publishing date, there are no established plans, budgetary or otherwise, to preserve the facilities at SLC-1 and SLC-2, nor are there any plans to establish a Thor Historic District at the National Park Service, Department of Interior or Air Force Space Command.

A young visitor smiles infectiously during his tour of SLC-10. The inclusion of history in a science, technology, engineering and mathematics (STEM) curriculum keeps the SLC-10 story relevant to future generations. *Courtesy of the author.*

Since the declassification of CORONA (1995), GAMBIT (2011), HEXAGON (2011) and QUILL (2012), it's no secret that Vandenberg launched the world's first photoreconnaissance satellites, due to its unique geography for launching into polar orbits. Vandenberg is the West Coast home of the NRO and, as such, should contain more than a handful of artifacts celebrating the organization's unique tale. Declassified Earthbound artifacts from the NRO are few and far between. GAMBIT and HEXAGON test articles reside at the National Museum of the U.S. Air Force in Dayton, Ohio. The sole remaining KH-4B CORONA camera resides at the National Air and Space Museum in Washington, D.C. No single location holds the bulk of the story of the early days of photoreconnaissance. Within a Thor Historic District, the location and artifacts will be able to tell a story to the NRO's employees about their heritage and legacy, while preserving the story for future generations.

An envisioned Thor Historic District would encompass the launch pads at SLC-1, SLC-2 and SLC-10 and remaining hardware and/or historic displays about:

- Royal Air Force participation in Project EMILY
- NRO utilization of Thor, with artifacts from CORONA
- Coverage of DMSP and Program 437
- The Delta family of boosters for NASA, NRO and commercial launches

The facilities at the SAMTEC already hold these artifacts, along with others from Atlas, Minuteman and Peacekeeper ICBMs. Christening of a Thor Historic District and the booster's unique place in space history would widen the outreach and impact to a larger segment of the interested population: space history aficionados, NRO personnel and distinguished visitors and dignitaries.

Appendix 1

OVERVIEW OF SLC-10 FACILITIES

Note: The entirety of SLC-10 consists of publicly accessible buildings and some that are off-limits. GPS coordinates are provided below as a means of identifying the structure's function, *not* tacit permission to go exploring on your own.

SLC-10 is classified as an industrial worksite, albeit deactivated, and contains many hazards you'd find at these locations: rusting metal, sharp objects, hazardous materials. DO NOT GO WANDERING AWAY FROM YOUR TOUR GROUP. Aside from an on-site Automated External Defibrillator and spartan first-aid kit, there are no medical facilities closer than twenty minutes away. Your tour guide will keep you safe, but remember to heed his/her directions when walking around the site.

Administration Building

N34.764275, W120.623040
The Admin building contains one-half of the Heritage Center collection, with artifacts from the CORONA program, donations from Brigadier General William King's estate and 10th Aerospace Defense Squadron items. The building also contains a retired Western Range console, which includes the "destruct" buttons for the Mission Flight Control Officer (MFCO) position. There is one shared-sex bathroom, no gift shop and no food/snack facililities. Any food or gift purchases must be made at the Base Exchange or in Lompoc/Santa Maria.

An aerial view of SLC-10 shows the West Pad (lower left), the Administration Building and Blockhouse (center) and Launch Emplacement 8 (upper left). The East Pad is not visible. *Courtesy of Vandenberg Space and Missile Technology Center.*

Technical Support Building (TSB) (aka "Museum Building")

N34.764079, W120.622871
The TSB was used to construct and repair support equipment components for Thor DMSP during SLC-10's launch activities in the mid- to late 1970s. Today, the building holds the lion's share of the Heritage Center artifacts. An Atlas-D launch console, commander and deputy consoles from a LGM-30F Minuteman II system and guidance "can" from a Minuteman II are just a few displays in the front half of the building. The Maintenance Bay holds Titan I Stage 1 rocket engines, an Agena mock-up and downstages from a Minuteman II. There is one shared-sex bathroom inside and a water fountain.

Public affairs officer Larry Hill (center left) gives a quick overview of satellite orbits to visitors inside the former Maintenance Bay. *Courtesy of the author.*

Photographers

The lighting environment inside the Admin building is similar to average office environment illumination (150–500 Lux). Some artifacts reside in glass cases, while others are displayed without. For the Museum Building, the lighting environment inside the first half of the building is dark (20–150 Lux) and may require a flash. The Maintenance Bay contains scoop lights, providing soft light on displays (100–200 Lux). The West Pad provides a split environment also—plenty of light as you enter and increasing darkness as you go toward the Thor booster (varies from 20 to 200 Lux).

SLC-10 West (aka "West Pad")

N34.763540, W120.624889

While the original launch equipment located at the West Pad was moved to JI to support Program 437, Douglas Aircraft Company rebuilt the site in 1963 from material returning from Project EMILY. Additional modifications, such as a "clean room" to support DMSP satellite launches, made SLC-10W one-of-a-kind.

The West Pad is located approximately nine hundred feet southwest of the Blockhouse. Access to the pad is provided via a sidewalk from the Admin building. The primary facility on the West Pad is the Launch Shelter (Facility 1658). Additional equipment supporting the missile during launch preparation resides around the shelter.

The sun sets over the Pacific Ocean, briefly illuminating the West Pad shelter at SLC-10. *Courtesy of the author.*

Launch Emplacement-8 (Facility 1661)

N34.766323, W120.622766

LE-8 is located approximately eight hundred feet northwest of the Blockhouse. It was the location of the first RAF IWST launch, code-named LIONS ROAR on April 16, 1959. The pad consisted of typical Thor missile shelter, propellant facilities and support facilities, identical to the East and West Pads. The Air Force dismantled LE-8 to support Operation DOMINIC in 1962. Facilities were later rebuilt at LE-8, but due to lack of use, the site was salvaged in the early 1970s. The remnants of LE-8 include the concrete blast walls, steel guide tracks used by the moveable shelter and shallow cable trenches. The site was a filming location for the ABC Family Channel drama *Rocket's Red Glare* in 2000. A fading 30th Space Wing logo, added for the film, can be seen on the blast walls.

Filming

Because SLC-10 is a National Historic Landmark and the Heritage Center tour is meant for public consumption, filming tour presentations for personal use is allowed. If your filming desires go into the realm of nonprofit usage, I would strongly recommend talking to the executive director and staff personnel. If you are interested in commercial filming at SLC-10, please contact the Air Force Entertainment Liaison Office at www.airforcehollywood.af.mil.

Today, LE-8 lies abandoned and stripped. The site of the RAF's first and last Thor launches has not yet been commemorated or celebrated by either country. *Courtesy of the author.*

OFF-LIMITS FACILITIES

SLC-10 East (aka "East Pad")

N34.762568, W120.621403
The East Pad is located approximately eight hundred feet south of the Blockhouse. The primary facilities on the East Pad were the launch shelter, propellant facilities and support facilities. The coastal climate has taken its toll on the launch shelter; it is considered a hazardous location due to the amount of rust and crumbling metal supports.

The Blockhouse

N34.764622, W120.621371
The Blockhouse, denoted in the singular but consisting of two conjoined but distinct buildings, contained the command and control equipment for the East and West Pads (in addition to LE-8), a distinguished visitor seating area and a Thor Technical Library. The Blockhouse, also known as the

The East Pad, unused since the shuttering of Program 437, falls prey to corrosion from the coastal climate. It is currently used for storage of excess museum artifacts. *Courtesy of the author.*

The SLC-10 Blockhouse underwent restoration during late 2015–early 2016. Note the Penthouse atop the structure. *Courtesy of the author.*

Launch Control Center or the Launch Operations Building, served as the main control station for SLC-10.

East Blockhouse[174]

Completed in 1959, the East Blockhouse was operated by the 392nd Missile Training Squadron as the operations and communications control center for launch activities on the East Pad. The one-story building contains approximately 1,600 square feet, measuring 52 feet wide by 43 feet in length. The building was made of "hardened" construction designed to protect technicians (as well as equipment) from the launch blast, in addition to any accidents or possible direct hits from a malfunctioning rocket. The reinforced concrete was spread through the foundation, floor, exterior walls and center walls. The flat roof has layers of reinforcing steel embedded in high-strength concrete.

Access into the East Blockhouse from the northeast side was through a five-foot, six-inch-wide single-panel steel blast door. The blast door was closed during launches and protected two sets of interior doors. The internal layout of the building was divided into four rooms along a central corridor. On the north side of the central corridor was the Communications Room, and a latrine and mechanical room were on the south side. The Launch Control Room measured approximately twenty feet wide by fifty feet in length, forming a T at the west end of the corridor. The control room was divided into equipment for the launch area on the north side and the telemetry equipment on the south side. Each area contained consoles, controls, equipment, monitors and TV stands specific to their designated missions. Aditionally, the Comm Room served as the communication hub for the entire SLC-10 complex.[175]

Nuclear Warriors and…Volleyball Champs?

In the Blockhouse parking lot, you'll notice a square border with holes in the ground for volleyball nets. According to 10th ADS member Eric Lemmon, due to the ninety-day rotations to JI, a large portion of the 10th ADS volleyball team remained at home. They became perennial winners in the Vandenberg volleyball tournaments. Lemmon quips, "Many of the opposing teams wondered why they kept seeing the same faces on the 10th ADS team, even after they had rotated to other bases for three years or so, and then returned to Vandenberg."[176]

West Blockhouse

Another change to the Blockhouse structure over the years, aside from the Penthouse, was a second Blockhouse addition called the SLC-10 West Blockhouse. Constructed in 1964, this blockhouse was built on the southwest side of the original structure as home to the 4300th Support Squadron. This structure was created as free-standing to maintain autonomy between the two squadrons. The center wall was actually two separate reinforced concrete walls, one for each blockhouse. Each blockhouse worked independent of each other (for training and operations) until 1968, when the two units were merged into the 10th ADS. By that time, a doorway was cut through the center concrete wall, and the West Blockhouse became the single control center for the complex.

Inside the West Blockhouse, the original launch support equipment still remains above a raised floor containing hundreds of feet of cabling. A note about launch operations inside the blockhouses: personnel did not have any direct line of sight to their respective launch pads, so TV monitors were installed in both, along with cameras throughout the complex to give complete situational awareness of the complex.

Prior Permission

Over the years, SLC-10 has supported researchers in a variety of fields. Two noteable groups were Project EMILY researchers from the United Kingdom and an academic doing a history on the evolution of rocket launch camera technology—the people who photographed/filmed the launches. Once permission was obtained, special dispensation was given for their fields of interest. Call the executive director at (805) 605-8300 if you are interested in doing research at SLC-10.

The Penthouse

Atop the Blockhouse is a small structure originally constructed as a study/relaxation area (break room) for RAF students. As a "self-help" project initated by 392nd MTS students, the Penthouse atop the Blockhouse was completed in 1962. The wood-framed structure was bolted to the flat roof to provide extra space for the students, since every inch of the Blockhouse was dedicated to the mission equipment. The structure was divided into two rooms totaling 814 square feet. It was constructed with two- by six-inch wood joists and rafters and two- by four-inch studs with cementitious panels on the exterior and gypsum sheetrock on the inside. Acoustic tiles covered the ceiling, and the floors had twelve-inch-square tan vinyl tiles. A steel staircase on the north side of the Blockhouse provided access to the Penthouse. After undergoing decades of disrepair, restoration funds were obtained in 2015. When restoration is complete, the rooms will contain a conference area and a STEM classroom for children and students. Contact SLC-10 personnel and ask if the Blockhouse or Penthouse is open for guided tours.

Surrounding Area

The rest of the SLC-10 off-road area is off-limits due to California and national wildlife restrictions. Many wildlife species reside in Vandenberg base perimeter, and some are protected by state and federal law. Any incursions into these protected areas will be reported to the Conservation Law Enforcement section of Vandenberg's Security Forces.

Appendix 2
LAUNCHES FROM SLC-10

TABLE 7. LAUNCHES FROM SLC-10/COMPLEX 75-2

Date	Pad	Launch Configuration	Launch Name	Program Association	Comments
16 Apr 59	75-2-8	Thor DM-18A (Initial Operational Capability program)	LIONS ROAR	Thor IRBM CTL IWST (RAF)	First Thor launched by RAF crew; successful
16 Jun 59	75-2-7	Thor DM-18A	RIFLE SHOT	Thor IRBM CTL IWST	RAF launch; failure
14 Aug 59	75-2-6	Thor DM-18A	SHORT SKIP	Thor IRBM; IOC	Failure
6 Oct 59	75-2-8	Thor DM-18A	FOREIGN TRAVEL	Thor IRBM; CTL WS-315A	RAF launch; successful
2 Mar 60	75-2-8	Thor DM-18A	CENTER BOARD	Thor IRBM; CTL WS-315A	RAF launch; successful
22 Jun 60	75-2-7	Thor DM-18A	CLAN CHATTAN	Thor IRBM; CTL WS-315A	RAF launch; first missile returned from UK for launching at VAFB; successful

Appendix 2

Date	Pad	Launch Configuration	Launch Name	Program Association	Comments
11 Oct 60	75-2-8	Thor DM-18A	LEFT RUDDER	Thor IRBM; CTL WS-315A	RAF launch; second missile returned from UK for launching at VAFB; successful
13 Dec 60	75-2-8	Thor DM-18A	ACTON TOWN	Thor IRBM; CTL WS-315A	RAF launch; successful
29 Mar 61	75-2-7	Thor DM-18A	SHEPHERDS BUSH	Thor IRBM; CTL WS-315A	RAF launch; successful
20 Jun 61	75-2-7	Thor DM-18A	WHITE BISHOP	Thor IRBM; CTL WS-315A	RAF launch; successful
6 Sep 61	LE-7	Thor DM-18A	SKYE BOAT	Thor IRBM; CTL WS-315A	RAF launch; successful
5 Dec 61	LE-8	Thor DM-18A	PIPERS DELIGHT	Thor IRBM; CTL WS-315A	RAF launch; successful
19 Mar 62	LE-7	Thor DM-18A	BLACK KNIFE	Thor IRBM; CTL WS-315A	RAF launch; failure
18 Jun 62	LE-8	Thor DM-18A	BLAZING CINDERS	Thor IRBM; CTL WS-315A	RAF launch; last CTL operational test launch; successful
19 Jan 65	4300 B-6	Thor-Burner I	ASTRAL LAMP	DMSP	Payload failed to separate from second stage
18 Mar 65	4300 B-6	Thor-Burner I	ASTRAL BODY	DMSP	Successful
20 May 65	4300 B-6	Thor Burner I	ROYAL EAGLE	DMSP	Successful
10 Sep 65	4300 B-6	Thor-Burner I	VICTORIA CROSS	DMSP	Successful
8 Jan 66	4300 B-6	Thor-Burner I	PERSIAN LAMB	DMSP	Second stage did not separate from Thor

Date	Pad	Launch Configuration	Launch Name	Program Association	Comments
31 Mar 66	4300 B-6	Thor-Burner I	RESORT HOTEL	DMSP	Last Thor-Altair (Burner I); successful
16 Sep 66	4300 B-6	Thor-Burner II	IRISH DUKE	DMSP	Successful
8 Feb 67	4300 B-6	Thor-Burner II	ARROW POINT	DMSP	Successful
29 Jun 67	LE-6	Thor-Burner II	DEER FOOT	*	Science, geodesy
23 Aug 67	LE-6	Thor-Burner II	*	DMSP	Successful
11 Oct 67	LE-6	Thor-Burner II	*	DMSP	Successful
23 May 68	SLC-10W	Thor-Burner II	*	DMSP	Successful
23 Oct 68	SLC-10W	Thor-Burner II	*	DMSP	Successful
23 Jul 69	SLC-10W	Thor-Burner II	*	DMSP	Successful
11 Feb 70	SLC-10W	Thor-Burner II	*	DMSP	Successful
3 Sep 70	SLC-10W	Thor-Burner II	*	DMSP	Successful
17 Feb 71	SLC-10W	Thor-Burner II	*	DMSP	Successful
8 Jun 71	SLC-10W	Thor-Burner II	*	*	Tested new celestial infrared sensors, spacecraft attitude sensing device
14 Oct 71	SLC-10W	Thor-Burner IIA	*	DMSP	Successful
24 Mar 72	SLC-10W	Thor-Burner IIA	*	DMSP	Successful
9 Nov 72	SLC-10W	Thor-Burner IIA	*	DMSP	Successful
17 Aug 73	SLC-10W	Thor-Burner IIA	*	DMSP	Successful
16 Mar 74	SLC-10W	Thor-Burner IIA	*	DMSP	Successful
9 Aug 74	SLC-10W	Thor-Burner IIA	*	DMSP	Successful
24 May 75	SLC-10W	Thor-Burner IIA	*	DMSP	Successful
19 Feb 76	SLC-10W	Thor-Burner IIA	*	DMSP	Failed to reach operational orbit
11 Sep 76	SLC-10W	Thor-Burner IIA	*	DMSP	Successful
1 May 78	SLC-10W	Thor-Burner IIA	*	DMSP	Successful
6 Jun 79	SLC-10W	Thor-Burner IIA	*	DMSP	Successful
15 Jul 80	SLC-10W	Thor-Burner IIA	*	DMSP	Failure

Appendix 3

THOR SPECIFICATIONS

TABLE 8. SPECIFICATIONS FOR THOR VARIANTS LAUNCHED AT SLC-10

	Thor (IRBM)	**SLV-2D**	**SLV-2F**	**SLV-2F**
Main Contractor	Douglas Aircraft Company	McDonnell-Douglas Astronautics Co.	McDonnell-Douglas Astronautics Co.	McDonnell-Douglas Astronautics Co.
Type	Booster	Single-stage, liquid fuel booster	Single-stage, liquid fuel booster	Single-stage, liquid fuel booster
Upper Stage	---	Burner I	Burner II	Burner IIA
Length	55.9 ft	55.9 ft; 6.0 ft (Burner I)	55.9 ft; 5.7 ft (Burner II)	55.9 ft; 6.25 ft (Burner IIA)
Diameter	8 ft	8 ft; 1.5 ft (Burner I)	8 ft; 5.4 ft (Burner II)	8 ft; 5.2 ft (Burner IIA)
Launch Weight	105,884 lbs	105,884 lbs; ? (Burner I)	105,884 lbs; 1,800 lbs (Burner II)	105,884 lbs; 2,400 lbs (Burner IIA)
Thrust	172,000 lbs	172,000 lbs; 5,000 lbs (Burner I)	172,000 lbs; 10,000 lbs (Burner II)	172,000 lbs; 10,000 lbs (first motor); 7,800 lbs (second motor)
Guidance	Inertial	Radio; Inertial (Burner II)	Radio; Inertial (Burner II)	Radio; Inertial (Burner IIA)
First/Last Launches	16 Dec 58/ 18 Jun 62	18 Jan 65/30 Mar 66	15 Sep 66/ 8 Jun 71	14 Oct 71/ 15 Jul 80

Appendix 4

LINEAGE AND HONORS

10TH AEROSPACE DEFENSE GROUP (ADC)

LINEAGE: Constituted 10th Aerospace Defense Group and activated on 1 Nov 1966. Organized on 1 Jan 1967. Discontinued and inactivated on 31 Dec 1971.

ASSIGNMENTS: 9th Aerospace Defense Division, 1 Jan 1967–1 Jul 1968. 14th Aerospace Force, 1 Jul 1968–31 Dec 1971.

STATIONS: Vandenberg AFB, California, 1 Jan 1967–31 Dec 1971.

MISSILES: Thor, 1966–1971.

DECORATIONS: Air Force Outstanding Unit Awards: 15 Nov 1963–15 Apr 1966; 1 Jul 1966–1 Jan 1967; 1 Jan–31 Dec 1975; 1 Jul 1976–1 Jun 1978.

EMBLEM: Approved 21 Nov 1963.

10TH AEROSPACE DEFENSE SQUADRON (ADC)

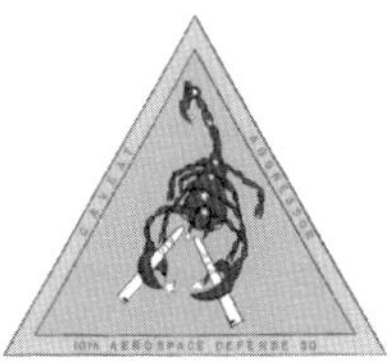

LINEAGE: Constituted 10th Aerospace Defense Squadron and activated on 24 Oct 1963. Organized on 15 Nov 1963. Discontinued and inactivated on 1 Jan 1967. Activated on 31 Dec 1970. Inactivated on 1 Nov 1979.

ASSIGNMENTS: Air Defense Command, 24 Oct 1963. 9th

Aerospace Defense Division, 1 Aug 1964–1 Jan 1967. 14th Aerospace Force, 31 Dec 1970. Aerospace Defense Command, 1 Oct 1976–1 Nov 1979.
STATIONS: Vandenberg AFB, California, 15 Nov 1963–1 Jan 1967. Vandenberg AFB, California, 31 Dec 1970–1 Nov 1979.
MISSILES: Thor, 1963–1979.
DECORATIONS: Air Force Outstanding Unit Awards: 15 Nov 1963–15 Apr 1966; 1 Jul 1966–1 Jan 1967; 1 Jan–31 Dec 1975; 1 Jul 1976–1 Jun 1978.

24TH AEROSPACE DEFENSE SQUADRON (ADC)

LINEAGE: Constituted as 24th Support Squadron and activated on 1 Nov 1966. Redesignated as 24th Aerospace Defense Squadron on 8 Jan 1970. Inactivated on 31 Dec 1970.
ASSIGNMENTS: 10th Aerospace Defense Group, 1 Jan 1967–31 Dec 1970.
STATIONS: Johnston Island AFB, Johnston Island, 1 Jan 1967–31 Dec 1970.
MISSILES: Thor, 1966–1970.
DECORATIONS: Air Force Outstanding Unit Awards: 1 Jan 1967–1 Jul 1968.

392ND MISSILE TRAINING SQUADRON (SAC)

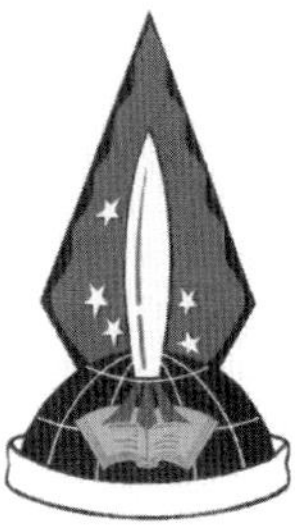

LINEAGE: Constituted as 392nd Missile Training Squadron on 20 May 1957. Activated on 15 Sep 1957. Redesignated as 392nd Missile Training Squadron (IRBM) on 1 Apr 1958. Discontinued and inactivated on 1 Feb 1963.
ASSIGNMENTS: 704th Strategic Missile Wing, 15 Sep 1957 (attached to 1st Missile Division, 6 Apr–30 Jun 1959); 1st Missile (later 1st Strategic Aerospace) Division, 1 Jul 1959; 392nd Strategic Missile Wing, 18 Oct 1961; 1st Strategic Aerospace Division, 19 Dec 1961–1 Feb 1963.
STATIONS: Cooke (later Vandenberg) AFB, California, 15 Sep 1957–1 Feb 1963.
MISSILES: Thor, 1958–1962.
OPERATIONS: Thor missile operations and maintenance training, primarily for Royal Air Force missile personnel, Aug 1958–Jun 1962.
DECORATIONS: Air Force Outstanding Unit Awards: 1 Jan 1958–30 Jun 1962.

4300th Support Squadron (SAC)

LINEAGE: Constituted as 4300th Support Squadron and organized on 1 Feb 1963. Inactivated and discontinued on 25 May 1967.
ASSIGNMENTS: 4000th Support Group, 1 February 1963–25 May 1967.
STATIONS: Vandenberg AFB, California, 1 Feb 1963–25 May 1967.
MISSILES: Thor, 1963–1967.
DECORATIONS: none.

394th Intercontinental Ballistic Missile Test Maintenance Squadron (SAC)

LINEAGE: Constituted as 394th Missile Training Squadron (ICBM) on 6 Mar 1958. Activated on 1 Apr 1958. Inactivated on 15 Dec 1958. Redesignated 394th Missile Training Squadron (ICBM-Minuteman) and activated on 10 Jun 1960. Redesignated 394th Strategic Missile Squadron (ICBM-Minuteman) on 1 Feb 1964; 394th Intercontinental Ballistic Missile Test Maintenance Squadron on 1 Jul 1976.
ASSIGNMENTS: 704th Strategic Missile Wing (ICBM), 1 Apr–15 Dec 1958. 1st Missile (later 1st Strategic Aerospace) Division, 1 Jul 1960; 392nd Strategic Missile Wing, 18 Oct 1961; 1st Strategic Aerospace Division (later Strategic Missile Center), 20 Dec 1961.
STATIONS: Cooke (later Vandenberg) AFB, California, 1 Apr–15 Dec 1958. Vandenberg AFB, California, 10 Jun 1960–1 Jul 1994.
MISSILES: Thor, 1979–1981. Minuteman, 1960–1994.

Page 163, top: The logo of the 10th Aerospace Defense Group incorporates the 10th ADS scorpion motif while demonstrating Program 437's protective nature for threats aimed at the United States. *Page 163, bottom*: The 10th Aerospace Defense Squadron logo displays a fearsome scorpion holding two Thor missiles, indicative of Program 437's dual launch capability. *Page 164*: The logo of the 392nd Missile Training Squadron illustrates that knowledge is the foundation for its celestial endeavors. *Page 165*: The 4300th Support Squadron motto highlights the eventual usage of DMSP data to assist in maintaining peaceful relations worldwide. *All courtesy of the Air Force Historical Research Agency.*

ACRONYMS

ABM: antiballistic missile
ADC: Air Defense Command/Aerospace Defense Command
ADG: Aerospace Defense Group
ADS: Aerospace Defense Squadron
AFB: Air Force base
AFBMD: Air Force Ballistic Missile Division
AFSC: Air Force Systems Command
AFWTR: Air Force Western Test Range
ALERTORD: alert order
AO: authentication officer
ARDC: Air Research and Development Command
ASAT: antisatellite
BOMARC: Boeing/Michigan Aeronautical Research Center
CCD: charge-coupled device
CIA: Central Intelligence Agency
CINCONAD: commander in chief, Continental Air Defense Command
CINCSAC: commander in chief, Strategic Air Command
CTL: Combat Training Launch
DMSP: Defense Meteorological Support Program
DNRO: director, National Reconnaissance Office
DOC: Department of Commerce
ELINT: electronic intelligence
EMP: electromagnetic pulse

FDE: force development evaluation
GE: General Electric
GPS: global positioning system
GWC: Global Weather Central
HAER: Historic American Engineering Record
ICBM: intercontinental ballistic missile
IMINT: imagery intelligence
IOC: initial operating capability
IRBM: intermediate-range ballistic missile
IWST: Integrated Weapon System Training
JI: Johnston Island
JTF: joint task force
KH: Keyhole
LE: launch emplacement
LOX: liquid oxygen
MATS: Military Air Transport Service
MOL: Manned Orbiting Laboratory
NASA: National Aeronautics and Space Administration
NHL: National Historic Landmark
NMFPA: Naval Missile Facility, Point Arguello
NOAA: National Oceanic and Atmospheric Administration
NPOESS: National Polar-orbiting Operational Environmental Satellite System
NRO: National Reconnaissance Office
NRP: National Reconniassance Program
NSA: National Security Agency
NTM: national technical means
OAO: Orbiting Astronomical Observatory
OCO: Orbiting Carbon Observatory
OPSEC: operations security
PALC: Point Arguello Launch Complex
PHOTINT: photographic intelligence
R&D: research and development
RAF: Royal Air Force
RCA: Radio Corporation of America
RP: rocket propellant
RSO: range safety officer
SAC: Strategic Air Command
SALT: Strategic Arms Limitation Talks/Treaty

SAMSO: Space and Missile Systems Organization
SAMTEC: Space and Missile Technology Center
SAMTO: Space and Missile Test Organization
SECDEF: secretary of defense
SENSINT: sensitive intelligence
SIGINT: signals intelligence
SLC: space launch complex
SLV: space launch vehicle
SPADATS: Space Detection and Tracking System
SPO: system program office
STC: Satellite Test Center
STEM: science, technology, engineering and mathematics
STG: Space Task Group
STS: Space Transportation System
TAT: Thrust Augmented Thor
TDY: temporary duty
TENCAP: Tactical Exploitation of National Capabilities
UK: United Kingdom
USACE: United States Army Corps of Engineers
USAF: United States Air Force
USSR: Union of Soviet Socialist Republics
WS: weapon system

NOTES

Chapter 1

1. Space and Missile Heritage Center, *Space Launch Complex 10*, 1.
2. Austerman, "Program 437," 33.
3. Space and Missile Heritage Center, *Space Launch Complex 10*, 6.
4. Ibid., 2.
5. Berger, "History of the 1st Strategic Aerospace Division,", 17.
6. Space and Missile Heritage Center, "Fact Sheet," 1.
7. Arms, "Thor," 1–2.
8. Space and Missile Heritage Center, *Space Launch Complex 10*, 5.

Chapter 2

9. Rockefeller, *History of Thor*, ch. 1.
10. Ibid., 4.
11. Ibid., 6.
12. Ibid., 8.
13. "Official Air Force Biography: Schriever."
14. U.S. Department of Defense, "DoD 4400.1-M," 8.
15. Rockefeller, *History of Thor*, 8.
16. Richards and Powell, "Waste Not," 1.
17. Sheehan, *Fiery Peace in a Cold War*, 318.
18. *Thor*, 863.
19. Technical Order 21-SM75-01, Missile and Equipment: SM-75 Weapon System, 1-2.

20. Rockefeller, *History of Thor*, 21.
21. Technical Order 21-SM75-01, Missile and Equipment: SM-75 Weapon System, 1-1.
22. Sheehan, *Fiery Peace in a Cold War*, 323.
23. Ibid., 333.
24. Rockefeller, *History of Thor*, 18.
25. Sheehan, *Fiery Peace in a Cold War*, 342.
26. Ibid.
27. Hunter, "Thor-Delta," 1.
28. BMDO, "Chronology of BMO," 334.
29. Berger, "History of the 1st Strategic Aerospace Division," 5.
30. Palmer, *Central Coast Continuum*, 121.
31. Berger, "History of the 1st Strategic Aerospace Division," 14.
32. National Park Service, SLC-10 HAER, 21.
33. Berger, "History of the 1st Strategic Aerospace Division, 1.
34. Hunter, "Thor-Delta," 1–8.

Chapter 3

35. NSA, Suez Crisis: A Brief COMINT History, 3.
36. Ibid., 16.
37. Ibid., 32.
38. Boyes, Thor Ballistic Missile, 38.
39. Hanner, "Chronology of the 392d Missile Training Squadron," ii.
40. Boyes, Project EMILY, 47.
41. Arms, "Thor," 2–11.
42. Boyes, Project EMILY, 65.
43. Arms, "Thor," 3–6.
44. Sheehan, Fiery Peace in a Cold War, 378.
45. Briggs, "Emily in Retrospect," 11.
46. Boyes, Thor Ballistic Missile, 61.
47. Arms, "Thor," 2–12.
48. Hanner, "Chronology of the 392d Missile Training Squadron," 2.
49. Boyes, Thor Ballistic Missile, 193.
50. Hanner, "Chronology of the 392d Missile Training Squadron," 2.
51. U.S. Army Engineer Research and Development Center, SLC-10 HAER, 22.
52. Ibid., 35.

53. Hanner, "Chronology of the 392d Missile Training Squadron," 19.
54. SAC, "SAC Missile Chronology," 23.
55. Ibid.," 34.
56. Briggs, "Emily in Retrospect," 2.

Chapter 4

57. DNA, "Operation DOMINIC I—1962," 26.
58. Plowshare Gnome in December 1961.
59. Now known as Sandia National Laboratories.
60. DTRA, "Fact Sheet: Operation DOMINIC I," 1.
61. Hansen, *U.S. Nuclear Weapons*, 82.
62. Austerman, "Program 437," 99.
63. Streeter, "Johnston Memories."
64. 48 U.S. Code, ch. 8.
65. Austerman, "Program 437," 99.
66. Ibid., 102.
67. DNA, "Operation DOMINIC I," 26.
68. *New York Times*, December 3, 1961.
69. Austerman, "Program 437," 103.
70. DNA, "Operation DOMINIC I," 87.
71. Ibid., 95.
72. Hansen, *U.S. Nuclear Weapons*, 87.
73. DNA, "Operation DOMINIC I," 228.
74. Vittitoe, "Did High-Altitude EMP Cause," 3.
75. DNA, "Operation DOMINIC I," 229.
76. Austerman, "Program 437," 13.
77. Peebles, *High Frontier*, 61.
78. Austerman, "Program 437," 53.
79. Temple, *Shades of Gray*, 361.
80. Austerman, "Program 437," 21.
81. Ibid., 25.
82. Chun, "Shooting Down a Star," 18.
83. Lemmon, July 30, 2015.
84. Johnston Atoll Safety Orientation, 7.
85. Johnston Island "deed."
86. Peebles, *High Frontier*, 59.
87. Molczan, "Program 437AP."

88. Austerman, *Program 437*, 30.
89. DAU, Glossary of Defense Acqusisition Acroynms and Terms.
90. Temple, *Shades of Gray*, 162.
91. Austerman, "Program 437," 53.
92. Peebles, *High Frontier*, 63.
93. Austerman, "Program 437," 58.
94. Molczan, "Program 437AP."
95. Chun, "Shooting Down a Star," 17.
96. Rottman, *World War II Pacific Island Guide*, 46.
97. Sandusky, "Hurricane Celeste Damage Assessment."
98. Carter, remarks.

Chapter 5

99. Oder, Fitzpatrick and Worthman, *GAMBIT Story*, 118.
100. Peebles, *CORONA Project*, 2.
101. R. Cargill Hall, "14 April 1956 Overflight of Noril'sk," 1.
102. Ibid., 3.
103. Peebles, *Guardians*, 10.
104. CIA, The Central Intelligence Agency and Overhead Reconnaissance, 85.
105. NRO Jr. website.
106. Directorate of Science & Technology. *History of the Office of Special Activities*, 102.
107. CIA, Central Intelligence Agency and Overhead Reconnaissance, 165.
108. Temple, *Shades of Gray*, 143.
109. Ruffner, *CORONA*, 14.
110. Oder, Fitzpatrick and Worthman, *CORONA Story*, 47.
111. Peebles, *CORONA Project*, 67.
112. NRO, Review and Redaction Guide, 2012 edition, 233.
113. Kiernan, "CIA's White Lies," June 3, 1995.
114. Oder, *CORONA Story*.
115. Peebles, *CORONA Project*, 68.
116. Oder, Fitzpatrick and Worthman, *CORONA Story*, 47.
117. CIA, *CORONA Program History*, vol. 1, 5-2.
118. Oder, *CORONA Story*, 48.
119. CIA, *CORONA Program History*, vol. 1, 5–7.
120. Ibid., 5–8.

121. NRO, Project AFTRACK records collection.
122. Peebles, *CORONA Project*, 128.
123. NRO, CAL Photograph Collection, Barcode 1400041672.
124. Ibid., 316.
125. Richelson, "Civilians, Spies and Blue Suits," 1.
126. Ibid., 86.
127. Oder, Fitzpatrick and Worthman, *GAMBIT Story*, 15.
128. Ibid., 14.
129. Ibid., 32.
130. Perry, *History of Satellite Reconnaissance*, vol. IIIA, 40.
131. Oder, Fitzpatrick and Worthman, *GAMBIT Story*, 41.
132. Ibid., 136.
133. McDonald and Wildlake, "Looking Closer," 3.
134. NRO, GAMBIT-1 Factsheet, 2.
135. Perry, *History of Satellite Reconnaissance*, vol. IIIB, 1.
136. Oder, Fitzpatrick and Worthman, *HEXAGON Story*, 65.
137. Ibid., 97.
138. Ibid., 116.
139. NRO, KH-10/DORIAN Manned/Unmanned Comparison Chart, 1.
140. NRO, QUILL records collection.
141. NRO, Cable Discontinuing QUILL DTG 102019Z FEB 69.
142. NRO, Review and Redaction Guide, 27.
143. Burrows, *Deep Black*, VII.

Chapter 6

144. NRO, Staff Records Collection.
145. NRO, "Hexagon (KH-9) Mapping Camera Program and Evolution," 99.
146. Mirroring usage in R. Cargill Hall's monograph due to the various program names throughout its history (Program 35, 698BH, 417), this book will use the term DMSP for consistency.
147. Boucher and Stier, "DMSP Instruments," 1.
148. Hall, *History of the Military Polar Orbiting Meteorological Satellite Program*, 16.
149. NRO, *HEXAGON Mapping Program*, 140.
150. Hall, *History of the Military Polar Orbiting Meteorological Satellite Program*, 10.
151. Ibid., 90.
152. Waltrop, "Underwater Ice Station Zebra," 5.
153. McCormack, "Rescue of Apollo 11," 1.

154. Mulcahy, CORONA Star Catchers.
155. Executive Order 12951, Release of Imagery Acquired by Space-Based National *Intelligence Reconnaissance Systems.* Only CORONA (KH-1,2,3,4A,4B), ARGON (KH-5) and LANYARD (KH-6) were declassified with this memorandum. Later systems were declassified in 2011.
156. McCormack, "Rescue of Apollo 11," 4.
157. Ibid.
158. John McLucas, Memorandum for Mr. Laird, Re: Taking Stock of the National Reconnaissance Program, NRO Collection, 1972, www.nro.gov/foia/declass/GAMHEX/HEXAGON/9.PDF.
159. MEMORANDUM FOR THE CHIEF OF STAFF, USAF. SUBJECT: Declassification of DSAP Data, 31 Aug 1972.
160. Hall, *History of the Military Polar Orbiting Meteorological Satellite Program*, 11.
161. Teets, Congressional Hearing Testimony.

Chapter 7

162. Temple, *Shades of Gray*, 481.
163. Kyle, "Thor Burner."
164. Western Test Range, Annual History 1980, 59.
165. NSDD 8, Space Transportation System.
166. U.S. Army Engineer Research and Development Center, SLC-10 HAER, 43.
167. Eleazer, "When 'About Time' Equals 'Too Late.'"
168. Air Force Space and Missile Museum, "Space Launch Complex 2W."
169. Department of the Interior, "National Register of Historic Places Inventory," 3.
170. Department of the Air Force, "Biography of Lieutenant General Donald O. Aldridge."

Chapter 9

171. Phone conversation with executive director Prichard, www.nrojr.gov/teamrecon/res_ba/articles/Art%201%20Rescue%20of%20Apollo%2011%20Rev4.pdf.
172. Phipps, "Space and Missile Heritage Center."
173. W6AB, "How the Satellite Amateur Radio Club Got Its Name."

Appendix 1

174. National Park Service, "Historic American Engineering Record," 51.
175. Ibid.
176. Lemmon, July 30, 2015.

BIBLIOGRAPHY

Books

Arnold, David C. *Spying from Space: Constructing America's Satellite Command and Control Systems*. College Station: Texas A&M University Press, 2005.

Boyes, John. *Project EMILY: Project Emily: Thor IRBM and the RAF*. Stroud, Gloucestershire, UK: The History Press, 2008.

———. *Thor Ballistic Missile: The United States and the United Kingdom in Partnership*. Havertown, PA: Fonthill Media, 2015.

Burrows, William. *Deep Black: Space Espionage and National Security*. New York: Random House, 1986.

Chun, Clayton. *Shooting Down a Star: The US Thor Program 437, Nuclear ASAT, and Copycat Killers*. Cadre Papers. Maxwell AFB, AL: Air University Press, 2000. handle.dtic.mil/100.2/ADA377346.

Del Papa, Michael. *From Snark to Peacekeeper: A Pictorial History of Strategic Air Command Missiles*. Omaha, NE: Office of the Historian, Strategic Air Command, 1990.

Gibson, James N. *Nuclear Weapons of the United States: An Illustrated History*. Atglen, PA: Schiffer Publishing, 1996.

Hall, R. Cargill. *A History of the Military Polar Orbiting Meteorological Satellite Program*. Office of the Historian. Chantilly, VA: National Reconnaissance Office, 2001.

Hansen, Chuck. *U.S. Nuclear Weapons: The Secret History*. Arlington, TX: Aerofax Inc., 1988.

Hartt, Julian. *The Mighty Thor: Missile in Readiness.* New York: Duell, Sloan and Pearce, 1961.

Lonnquest, John, and David Winkler. *To Defend and Deter: The Legacy of the United States Cold War Missile Program*. Washington, D.C.: Department of Defense Legacy Resource Management Program, 1996.

Mulcahy, Robert, Jr., ed. *CORONA Star Catchers: The Air Force Aerial Recovery Aircrews of the 6593rd Test Squadron (Special), 1958–1972*. Washington, D.C.: U.S. Government Printing Office, 2012.

Norris, Pat. *Spies in the Sky: Surveillance Satellites in War and Peace*. New York: Springer Publishing, 2008.

Oder, Frederic, James Fitzpatrick and Paul Worthman. *The CORONA Story*. Washington, D.C.: U.S. Government Printing Office, 1988.

———. *The GAMBIT Story*. Washington, D.C.: U.S. Government Printing Office, 1988.

———. *Guardians: Strategic Reconnaissance Satellites.* New York: Random House, 1987.

———. *The HEXAGON Story*. Washington, D.C.: U.S. Government Printing Office, 1988.

Peebles, Curtis. *The CORONA Project: America's First Spy Satellites*. Annapolis, MD: Naval Institute Press, 1997.

———. *High Frontier: The United States Air Force and the Military Space Program*. Washington, D.C.: Air Force History and Museums Program, 1997.

Perry, Robert. *History of Satellite Reconnaissance*. Vol. IIIA, *GAMBIT*. Chantilly, VA: National Reconnaissance Office.

———. *History of Satellite Reconnaissance*. Vol. IIIB, *HEXAGON*. Chantilly, VA: National Reconnaissance Office.

Polmar, Norman, and Robert S. Norris. *The U.S. Nuclear Arsenal: A History of Weapons and Delivery Systems since 1945*. Annapolis, MD: Naval Institute Press, 2009.

Rottman, Gordon L. *World War II Pacific Island Guide: A Geo-Military Study*. Westport, CT: Greenwood Press, 2002.

Ruffner, Kevin C. *CORONA: America's First Satellite Program*. Langley, VA: Central Intelligence Agency, 1995.

Sheehan, Neil. *A Fiery Peace in a Cold War: Bernard Schriever and the Ultimate Weapon*. New York: Random House, 2009.

Spires, David. *On Alert: An Operational History of the United States Air Force Intercontinental Ballistic Missile Program, 1945–2011*. Colorado Springs, CO: Air Force Space Command History Office, 2012.

Temple, L. Parker, III. *Shades of Gray: National Security and the Evolution of Space Reconnaissance*. Reston, VA: American Institute of Aeronautics and Astronautics, 2005.

Newspapers

Phipps, Shane. "Space and Missile Heritage Center Preserves Past to Conserve Future." 30th Space Wing Public Affairs. February 13, 2015. www.vandenberg.af.mil/news/story.asp?id=123439223.

Pamphlets

From Tanks to Missiles: Vandenberg Air Force Base and the 30th Space Wing from Camp Cooke to the Present. 30th Space Wing History Office, 1995.

McDonald, Robert A., and Sharon K. Moreno. *Raising the Periscope: GRAB and POPPY…America's Early ELINT Satellites*. Center for the Study of National Reconnaissance. Chantilly, VA: National Reconnaissance Office, 2005. www.nro.gov/history/csnr/programs/docs/prog-hist-03.pdf.

Space and Missile Heritage Center. *Space Launch Complex 10: A National Historic Landmark*. Vandenberg AFB, CA: 30th Civil Engineering Squadron Cultural Affairs, n.d.

Articles

Boucher, Donald, and Anthony Stier. "DMSP Instruments: A 50-Year Legacy." *Crosslink*, Spring 2010. El Segundo, CA: Aerospace Corporation, 2010. www.aerospace.org/crosslinkmag/spring-2010/dmsp-instruments-a-50-year-legacy.

Briggs, R.S., ed. "Emily in Retrospect." *Airview News*. Santa Monica, CA: Douglas Aircraft Company, April 1960.

Hall, R. Cargill. "The 14 April 1956 Overflight of Noril'sk, U.S.S.R." Chantilly, VA: National Reconnaissance Office, 2003. www.nro.gov/foia/declass/Archive/15-02.PDF.

Kiernan, Vincent. "CIA's White Lies over Mice in Space." *New Scientist*, June 3, 1995. www.newscientist.com/article/mg14619800-600-cias-white-lies-over-mice-in-space.

McCormack, Noel. "The Rescue of Apollo 11: Corona and DMSP's Unforeseen Mission." Center for the Study of National Reconnaissance, 2005. www.nrojr.gov/teamrecon/res_ba/articles/Art%201%20Rescue%20of%20Apollo%2011%20Rev4.pdf.

McDonald, Robert A., and Patrick Widlake. "Looking Closer and Looking Broader: Gambit and Hexagon—The Peak of Film-Return Space Reconnaissance after Corona." *National Reconnaissance Journal of the Discipline and Practice*. Chantilly, VA: Center for the Study of National Reconnaissance, 2012.

Molczan, Ted. "Program 437AP: A Sub-orbital Corona-Derived Satellite Inspector." 2016.

Ray, Justin. "What to Do with the Final Delta 2 Rocket?" *Spaceflight Now*, April 8, 2015. spaceflightnow.com/2015/04/08/what-to-do-with-the-final-delta-2-rocket.

Richards, G.R., and J.W. Powell. "Waste Not—The Use of Ex-RAF Thor Vehicles." *Journal of the British Interplanentary Society* 50 (1997): 189–200.

Waltrop, David. "An Underwater Ice Station Zebra: Recovering a Secret Spy Capsule from 16,400 Feet Below the Pacific Ocean." Langley, VA: Central Intelligence Agency, 2012. www.cia.gov/library/publications/cold-war/underwater-ice-station-zebra/ice-station-zebra.pdf.

Memoranda

Defense Acquisition University. "Glossary of Defense Acquisition Acronyms and Terms." Washington, D.C. Defense Acqusisition University, 2016. dap.dau.mil/glossary/pages/2747.aspx.

Department of the Interior. "National Register of Historic Places Inventory—Nomination Form: Space Launch Complex 10." focus.nps.gov/pdfhost/docs/NHLS/Text/86003511.pdf.

Executive Order 12951. "Release of Imagery Acquired by Space-Based National Intelligence Reconnaissance Systems." February 22, 1995, *Code of Federal Regulations*, title 3 (1995): 10789–790. www.gpo.gov/fdsys/pkg/FR-1995-02-28/pdf/95-5050.pdf.

48 U.S. Code Chapter 8—GUANO ISLANDS ACT. www.law.cornell.edu/uscode/text/48/chapter-8.

National Reconnaissance Office. "MEMORANDUM FOR THE CHIEF OF STAFF, USAF. SUBJECT: Declassification of DSAP Data." August 31, 1972.

———. "Review and Redaction Guide, 2012 edition." Chantilly, VA: National Reconnaissance Office, 2012.

National Security Decision Directive 8. *Space Transportation System* (1981). marshall.wpengine.com/wp-content/uploads/2013/09/NSDD-8-Space-Transportation-System-13-Nov-1981.pdf.

Sandusky, Vernon. "Hurricane Celeste Damage Assessment at Johnston Island." October 1972.

Letters and E-mails

All letters/e-mails are to the author unless otherwise noted.

Eleazer, Wayne. July 29, 2015.

Lemmon, Eric. Thor Association. July 30, 2015.

Reports

Arms, W.M. "Thor: The Workhorse of Space—A Narrative History." Huntington Beach, CA: McDonnell Douglas Astronautics Company–West, 1972.

Austerman, Wayne. "Program 437: The Air Force's First Antisatellite System." Colorado Springs, CO: Air Force Space Command History Office, 1991.

Berger, Carl. "History of the 1st Strategic Aerospace Division and Vandenberg Air Force Base 1957–1961." Vandenberg AFB, CA: Headquarters, 1st Strategic Aerospace Division, 1962.

Central Intelligence Agency. *The Central Intelligence Agency and Overhead Reconnaissance.* Office of History. Langley, VA: Central Intelligence Agency, 1988.

———. "CORONA Program History, Volume 1." Langley, VA: Central Intelligence Agency, 1972.

Defense Nuclear Agency. "Operation DOMINIC I—1962." Washington, D.C.: Defense Nuclear Agency, 1983. www.dtic.mil/dtic/tr/fulltext/u2/a136820.pdf.

Defense Threat Reduction Agency. "Fact Sheet: Operation DOMINIC I." www.dtra.mil/Portals/61/Documents/NTPR/1-Fact_Sheets/23_DOMINIC_I.pdf.

Directorate of Science & Technology. *History of the Office of Special Activities (OSA) from Inception to 1969*. Langley, VA: Central Intelligence Agency, 1969.

Hunter, Peter. "Thor-Delta: Launches 1957–2004." Sydney, AUS, n.d.

National Park Service. "Historic American Engineering Record of Space Launch Complex 10, Vandenberg Air Force Base, California." Omaha, NE: NPS Midwest Regional Office.

National Reconnaissance Office. "HEXAGON Mapping Program." Chantilly, VA: National Reconnaissance Office, 1982.

National Security Agency. "The Suez Crisis: A Brief COMINT History." Vol. 2. United States Cryptologic History, Special Series Crisis Collection. Fort Meade, MD: National Security Agency, 1988. www.nsa.gov/news-features/declassified-documents/cryptologic-histories/assets/files/Suez_Crisis.pdf.

Palmer, Kevin. *Central Coast Continuum—From Ranchos to Rockets: A Historic Overview for an Inventory and Evaluation of Historic Sites, Buildings, and Structures, Vandenberg Air Force Base, California.* Santa Maria, CA: BTG, Inc., 1999.

Rockefeller, Alfred. *History of Thor 1955–1959.* Los Angeles: Air Force Ballistic Missile Division, 1960.

"Thor: A Study of a Great Weapon System." Office of History. Los Angeles AFB, CA: Office of History, 1972.

Vittitoe, Charles. "Did High-Altitude EMP Cause the Hawaiian Streetlight Incident?" Albuquerque, NM: Sandia National Laboratories, 1989. www.ece.unm.edu/summa/notes/SDAN/0031.pdf.

Chronologies

Ballistic Missile Defense Office. "Chronology of Ballistic Missile Office 1945–1990." Washington, D.C., 1991.

1st Missile Division. Chronology of Events—1959.

Hanner, Ray. "Chronology of the 392d Missile Training Squadron (THOR)." Vandenberg AFB, CA: Headquarters, 1st Strategic Aerospace Division, 1961.

Strategic Air Command. "SAC Missile Chronology, 1939–1988." Offutt AFB, NE: Office of the Historian, 1989.

Western Test Range, Annual History, Fiscal Year 1980. Vandenberg AFB, CA: Office of the Historian, 1980.

Manuals

U.S. Department of Defense. "DoD 4400.1-M: Priorities and Allocations Manual." Washington, D.C.: Government Printing Office, 2002. www.dtic.mil/whs/directives/corres/pdf/440001m.pdf.

Speeches

Carter, Jimmy. Remarks at the Congressional Space Medal of Honor Awards Ceremony, Kennedy Space Center, Florida, October 1, 1978.

Teets, Peter. Congressional Hearing Testimony for the Undersecretary of the Air Force, the Honorable Peter B. Teets. February 25, 2004. www.nro.gov/news/testimony/2004/2004-02.pdf.

Collections

National Reconnaissance Office. CORONA, ARGON, LANYARD (CAL) Collection. Declassified November 26, 1997. www.nro.gov/foia/LibraryListingSQL.aspx.

———. GAMBIT and HEXAGON Records and Histories. Declassified September 17, 2011. www.nro.gov/foia/declass/GAMBHEX.html.

———. GAMBIT Dual Mode Records. Declassified January 10, 2013. www.nro.gov/foia/declass/GAMBIT%20Dual%20Mode.html.

———. HEXAGON 1201-3 Recovery Records. Declassified August 16, 2012. www.nro.gov/foia/declass/HEXAGON%20Recovery.html.

———. KH-9 HEXAGON Records. Declassified October 1, 2012. www.nro.gov/foia/declass/HEXAGON%20Records.html.

———. KH-10 DORIAN/Manned Orbiting Laboratory (MOL) Records. Declassified October 22, 2015. www.nro.gov/foia/declass/MOL.html.

———. NRO Staff Records. www.nro.gov/foia/declass/NROStaffRecords.html.

———. Project AFTRACK Records. Declassified October 7, 2015. www.nro.gov/foia/declass/AFTRACK.html.

———. QUILL Records. Declassified July 9, 2012. www.nro.gov/foia/declass/QUILL.html.

———. WS117L, SAMOS, and SENTRY Records. www.nro.gov/foia/declass/WS117L_Records.html.

Directories

Vandenberg Air Force Base phonebook.

Internet Sources

Air Force Space and Missile Museum. "Space Launch Complex 2W." afspacemuseum.org/vandenberg/SLC2W.

Department of the Air Force. "Biography of Lieutenant General Donald O. Aldridge." www.af.mil/AboutUs/Biographies/Display/tabid/225/Article/107864/lieutenant-general-donald-o-aldridge.aspx.

Eleazer, Wayne. "When 'About Time' Equals 'Too Late.'" *The Space Review*. October 11, 2005. www.thespacereview.com/article/470/1.

Jacobson, Willis. "Early Space Exploration Artifacts on Display at VAFB." *Santa Maria Times*, Feburary 27, 2015. santamariatimes.com/lompoc/news/local/early-space-exploration-artifacts-on-display-at-vafb/article_04712682-1fcd-5df8-8394-2bc50c68f115.html.

Kyle, Ed. "Thor Burner: Sixth in a Series Reviewing Thor Family History." February 27, 2011. www.spacelaunchreport.com/thorh6.html.

National Reconnaissance Office. KH-10/DORIAN Manned and Unmanned Comparison Charts. www.nro.gov/foia/declass/DORIAN/15-MANNED_UNMANNED_COMPARISON.PDF.

———. "NRO Jr. Webpage." www.nrojr.gov.

"Official Air Force Biography: General Bernard Adolph Schriever." www.af.mil/AboutUs/Biographies/Display/tabid/225/Article/104877/general-bernard-adolph-schriever.aspx.

Space and Missile Heritage Center. "Fact Sheet: Vandenberg Space and Missile Heritage Center." January 26, 2015. www.vandenberg.af.mil/library/factsheets/factsheet.asp?id=4627.

Streeter, Robert. "Johnston Memories." johnstonmemories.com.

W6AB. "How The Satellite Amateur Radio Club Got Its Name." Vandenberg Amateur Radio Club. www.satellitearc.com/clubname.html.

INDEX

A

ABM 63, 64
Aerospace Defense Command 67, 74
Agena 72, 82, 83, 84, 85, 86, 90, 91, 106, 122, 127, 148
Air Research and Development Command 25, 33
Apollo 11 109, 110, 111
Army 25, 27, 32, 34, 37, 64, 77, 95
ASAT 62, 64, 72, 74
Atlantic Ocean 30

B

Blockhouse 67, 138, 149, 150, 151, 153, 154
BLUEGILL 60, 61
BLUEGILL DOUBLE PRIME 61, 63
BLUEGILL PRIME 60, 61, 62
BLUEGILL TRIPLE PRIME 63

C

Caspian Sea Monster 92
CIA 79, 80, 82, 84, 85, 86, 87, 88, 92, 93
Cooke AFB 33, 36, 46
CORONA 17, 71, 75, 82, 83, 84, 85, 86, 87, 88, 89, 90, 91, 92, 93, 95, 97, 99, 105, 106, 110, 111, 115, 128, 141, 144, 147
CTLs 48
Cuban Missile Crisis 27, 62, 65, 105

D

Defense Meteorological Support Program (DMSP) 16, 18, 52, 69, 73, 75, 95, 101, 102, 105, 106, 107, 108, 109, 110, 111, 112, 113, 114, 115, 116, 118, 120, 127, 144, 149
Delta 9, 116, 121, 122, 129, 141, 143, 144

Discoverer 18, 19, 82, 83, 84, 85, 86, 141
DOMINIC 53, 54, 55, 58, 59, 60, 61, 62, 63, 122, 150
DORIAN 94

E

Eisenhower, Dwight 25, 43, 78
EMILY. See Project EMILY
EMP 61, 62

F

FISHBOWL 59, 60, 62, 63
437 AP 71, 72, 128
4300th Support Squadron 66, 154

G

GAMBIT 75, 88, 89, 90, 91, 92, 93, 96, 97, 110, 144
GAMBIT-1 90, 91, 92, 97
GAMBIT-3 90, 91, 92, 97
GENETRIX 78, 82
Global Weather Central 105, 106, 113

H

Hawaii 57, 58, 61, 74, 110, 111
Heritage Center 9, 124, 127, 137, 147, 148, 150
HEXAGON 75, 88, 92, 93, 94, 95, 96, 97, 107, 108, 109, 110, 144

I

ICBM 23, 26, 50, 53, 79, 80, 81, 90, 120, 127, 128, 131
Intelligence Community 77, 89, 113, 118
IRBM 15, 18, 23, 25, 27, 34, 36, 37, 41, 50, 71, 77, 122, 129
IWST 46, 48, 49, 50, 150

J

Johnston Island (JI) 10, 15, 55, 56, 57, 58, 59, 60, 62, 63, 65, 66, 67, 68, 69, 71, 72, 73, 74, 149, 153
Jupiter 25, 27, 37

K

KH-7 90, 92, 97
KH-8 90, 91, 92, 97
KH-10 94
KH-11 79, 96, 97
KINGFISH 62, 63
King Kamehameha 57, 58

L

Launch Emplacement 1 60
Launch Emplacement 8 (LE-8) 15

M

Manned Orbiting Laboratory 73, 94
McNamara, Robert 50, 64, 65, 66
Minuteman 25, 52, 55, 127, 128, 145, 148

N

National Aeronautics and Space Administration (NASA) 72, 100, 105, 110, 111, 116, 119, 120, 121, 128, 144
National Historic Landmark (NHL) 122, 124, 141, 143
National Oceanic and Atmospheric Administration (NOAA) 112, 113, 118, 141
National Park Service 18, 122, 144
National Reconnaissance Office (NRO) 83, 84, 86, 87, 88, 89, 100, 105, 107, 109, 111, 112, 113, 119, 131, 144, 145
National Reconnaissance Program (NRP) 75, 88, 95, 100, 105, 107, 110, 111
National Security Agency 38, 50
National Technical Means (NTM) 74
nuclear testing 15, 53, 56, 63

P

Pacific Ocean 18, 32, 49, 55, 56, 65, 69, 83, 86, 93, 109, 110
Pearl Harbor 58, 75, 76, 77, 111
Program 437 10, 15, 18, 52, 56, 62, 63, 65, 66, 67, 68, 69, 70, 71, 73, 74, 127, 144, 149
Program 505 64
Project EMILY 15, 37, 41, 50, 52, 66, 149, 154
Project MUDFLAP 64

Q

QUILL 95, 96, 144

R

RAF 15, 34, 40, 41, 43, 45, 46, 48, 49, 50, 60, 77, 122, 150, 154
RAF Lakenheath 43
RAND Corporation 78
Reagan, Ronald 75, 120
Rocketdyne 27, 30
Royal Air Force 15, 34, 46, 122, 144

S

SAC 33, 36, 40, 46, 66, 77, 95, 105, 107, 137
SAINT 71
SAMOS 90, 96
SAMTEC. See Space and Missile Technology Center
Satellite Test Center 108
SENSINT 78
SIGINT 39, 77, 86
SLC-10 9, 10, 15, 16, 18, 34, 46, 48, 62, 65, 67, 69, 95, 106, 118, 120, 122, 124, 127, 128, 129, 131, 137, 138, 141, 144, 147, 148, 149, 150, 151, 153, 154, 155, 157
SLC-10W 16, 60, 120, 149
Soviet Union 17, 25, 37, 54, 55, 63, 64, 65, 74, 76, 78, 80, 86, 91, 97, 99, 108, 110
Space and Missile Heritage Center 18, 127, 137
Space and Missile Technology Center 137

Sputnik 46, 53, 79, 82
SQUANTO TERROR 67
STARFISH 60
STARFISH PRIME 60, 61, 62, 65
STEM 137, 138, 139, 141, 155

T

10th ADG 73
10th ADS 65, 68, 69, 73, 74, 153, 154
Thor 15, 18, 23, 26, 27, 29, 30, 32, 34, 36, 37, 39, 40, 41, 43, 45, 46, 49, 50, 52, 55, 59, 60, 61, 62, 63, 64, 65, 66, 68, 69, 70, 71, 72, 74, 77, 82, 83, 85, 90, 105, 106, 115, 116, 119, 120, 121, 122, 127, 129, 137, 141, 143, 144, 145, 148, 149, 150, 151
  IRBM 25, 36, 55, 59, 67
392nd Missile Training Squadron 46, 153
TIGERFISH 60
TIROS 19, 100, 104
Titan 9, 26, 37, 50, 91, 116, 122, 127, 128, 148

U

U-2 78, 79, 80, 81, 82, 87, 88, 89
U.S. Air Force (USAF) 23, 46
U.S. Navy 58, 60, 86, 111

V

Vandenberg 9, 15, 18, 32, 34, 35, 36, 65, 67, 82, 127, 141, 143, 144, 153, 155
Vandenberg AFB 31, 33, 46, 50, 73, 75, 84, 86, 90, 96, 97, 105, 122, 137
Vietnam 73, 94, 107, 110

W

weather 16, 19, 89, 95, 97, 99, 100, 102, 105, 106, 107, 108, 109, 110, 111, 112, 113, 114, 116, 141
WS-117L 81, 82, 90, 95, 96, 99

ABOUT THE AUTHOR

Joseph T. "Joe" Page II is an amateur space historian and former Air Force space and missile officer. The space and history "bug" bit him early in life while growing up at White Sands Missile Range, New Mexico, the "Birthplace of America's Missile and Space Activity." After spending tours of duty in California, North Dakota and Afghanistan, Joe and his family finally settled in New Mexico, the "Land of Enchantment" and one of the few places in the civilized world where the light from the Milky Way creates shadows in the desert night.

Joe is the author of three titles in the Images of America series focusing on military aviation and space history. He holds a bachelor's of science degree in engineering technology from New Mexico State University and a master's of science in space studies from American Military University.